"As a person who is more than a decade into my transition and who doesn't identify as particularly spiritual, I was surprised by how instantly I felt seen by this book. Eli has created a beautiful guide to embracing rituals and meets the reader right where they are. Like our transitions, the navigation of this book can be catered to your own needs. I was struck by the realization that rituals are already present in my life, and gifted a more intentional framework through which to recognize and embrace them. As someone who is in a constant state of self-reflection, I love when I am offered the opportunity to think about gender and myself in new ways. *Gender Rituals* gave me that and so much more."

—MAYBE BURKE, artist, educator, and coauthor of *I Heart Orgasms*

"In a landscape where many of us daily navigate so many messages counter to our value and magic, Eli Lawliet's *Gender Rituals* is a must-read for anyone desiring to tune into the fullness of who we really are. Offering both historical context and specific how-to guide to ritual, this book is true spiritual medicine and is the culmination of a lineage of collective knowledge."

—TOURMALINE, artist, filmmaker, activist, and author of *Marsha: The Joy and Defiance of Marsha P. Johnson*

"There is nothing else out there like *Gender Rituals*. Each page illuminates the amount of thought, intention, research, and care that Lawliet has placed in this accessible modern-day masterpiece. I want everyone to read this book—regardless of their gender. In a world where we are constantly told who to be, *Gender Rituals* invites readers to ritualize their relationship with self, deepening their embodied presence in increasingly uncertain times. We are in immediate need of greater empathy across the planet. Readers can support the journey to a more ideal future for all of us, simply by starting with empathy for themselves. A better world is possible, and it's clear that Lawliet believes it."

— MARIA MINNIS, author of *Tarot for the Hard Work*

"In *Gender Rituals*, Eli Lawliet has composed an intelligently accessible and nourishing ritual primer. While prioritizing the needs of transfolk, this is a text that will support anyone—truly!—who is interested in well-considered ritual toward more honest and more loving human-beingness. Eli deftly provisions ritual guidance and encouragement without ever devolving into dogma, making room for generative exploration within well-considered guardrails. *Gender Rituals* is destined to become a well-loved anchor and compass, and a necessary addition to any gender-expansive magical library."

—DIANA ROSE, animist astrologer and coauthor of *The Rosebud Tarot*

Praise for *Gender Rituals*

"*Gender Rituals* is a revelation—a radical and wild benediction of a book. I gasped, wept, laughed, and nodded along the entire time I was reading. It answered so many unasked questions and met so many unmet longings I've carried for ritual, ceremony, prayer, and belonging across the many spirals of my own gender identity. It is profoundly affirming, supportive, and life-giving—just like the human who brought it into being."

— LINDSAY MACK, author and founder of *Tarot for the Wild Soul*

"*Gender Rituals* takes care of one of the most pertinent questions of being trans today: how to make more time, in the form of habits, patience, and history, for one another. Go read this book. It is both hands-on and gentle. In it, Dr. Lawliet teaches you how to be available for all kinds of things: opening and closing a container, grounding yourself where you are at, waiting for trust, and being a friend. Ritual-making is a very important skill for anyone who is trying to make loving community today and *Gender Rituals* is a great teacher for you if you are trying to bring more people together."

—ALICE SPARKLY KAT, author of *Postcolonial Astrology*

"Eli is as heart-centered as he is brilliant. In *Gender Rituals*, a book as comprehensive and grounded as it is innovative, he generously makes his expertise, honed over many years and hundreds of clients, available to anyone needing support."

— SARAH FAITH GOTTESDIENER, author of *The Moon Book*

"Eli Lawliet has created a beautiful book in *Gender Rituals*. As a lifelong magician and witch-thing who came to fully accept their non-binary nature late in life, I hadn't previously applied much ritual to this awareness. This book changed that, and I am very grateful for it. *Gender Rituals* will be especially helpful to various trans folks, but the body connection rituals, protection work, and ritual skills taught will benefit almost everyone. I found the practices uplifting, healing, and powerful. The naming rituals are particularly potent for those using a new, chosen name for any reason, as are those for releasing old narratives around identity that no longer serve us. There's a lot here for anyone, regardless of gender. Highly recommended!"

—AIDAN WACHTER, author of *Six Ways*, *Weaving Fate*, and *Changeling*

"*Gender Rituals* is not only a vital and excellent guide for anyone who wants to bring intention and spirit to their gender journey, but also a powerful blessing for our collective liberation. If you're interested in deepening your relationship to yourself, your gender, and our gender-blessed ancestors, you'll find an incomparable treasure trove of information and wise counsel in Eli Lawliet's inspired words. As a therapist and educator, I'm so grateful this timely resource has been brought into the world so skillfully and beautifully!"

—ALEX IANTAFFI, PhD, SEP, CST, LMFT, award-winning author of *Gender Trauma* and host of the *Gender Stories* podcast

"*Gender Rituals* is a beautiful love letter to trans and queer community, but it's so much more than that, too. The rituals crafted herein will alter your life for the better every single time you perform one—and if you let them, they'll alter *you* in the best way possible too. Eli's work has meant so many things to so many people over the years, and here it is both condensed down to an easy access point for wherever you are in your journey *and* yet still as expansive as the cosmos that we live under."

—CASSANDRA SNOW, author of *Queering the Tarot*, *Queering Your Craft*, coauthor of *Lessons From the Empress*, and editor of *Tarot In Other Words*

Gender Rituals

Creating a Sacred Space for Your Body, Your Gender, and Your Becoming

gender rituals

eli lawliet, phd

North Atlantic Books
Huichin, unceded Ohlone land
Berkeley, California

North Atlantic Books
An educational nonprofit organization
Huichin, unceded Ohlone land
2526 Martin Luther King Jr Way
Berkeley, CA 94704 USA
www.northatlanticbooks.com

Cover design by Jasmine Hromjak
Book design by Happenstance Type-O-Rama

Acquisitions and production editor: Gillian Hamel • Copyeditor: Christopher Church
Proofreader: Melanie Demmer • Indexer: Tim Tate

North Atlantic Books's publications are distributed to the US trade and internationally by Penguin Random House Publisher Services. For further information, visit our website at www.northatlanticbooks.com.

The authorized representative in the EU for product safety and compliance is Eucomply OÜ, Pärnu mnt 139b-14, 11317 Tallinn, Estonia, hello@eucompliancepartner.com, +33757690241.

Library of Congress Cataloging-in-Publication data is available from the publisher upon request.
ISBN: 979-8-88984-325-2 (paperback)
ISBN: 979-8-88984-326-9 (ebook)

1 2 3 4 5 6 7 8 9 KPC 30 29 28 27 26

Printed in the United States of America

For the transcestors, who insisted
I write this book now.
You were right, as always.
Thank you.

Contents

Acknowledgments

Nothing in this entire world happens in a vacuum, and no one writes a book alone. That said, I feel deeply that this book came from many sources, and that I was mostly a steward and a conduit for whatever needed to be transmitted. So first and foremost, my gratitude and acknowledgment to the transcestors, who poured this book into and through me. Thank you for trusting me with these words.

Next, my deepest acknowledgment and gratitude to the spirits who work with me, who hold me and protect me, who guide me and support me. My deepest acknowledgment to the land that I live on and the web of life and connection wherein I exist. My deepest gratitude to every life that is mingled with my own, and to the interlocking systems of support that hold me. Thank you for bringing me everything I needed to birth this work into the world.

My love and gratitude to my beloveds: Joon Seo, Zephyr, Bonnie, Gabe, Zel, and Mars. Each of you is represented in various parts of this work, and each of you has provided me with irreplaceable support and care throughout the process of writing it.

Bonnie, thank you for writing with me in the desert, for enjoying our Midwestern party comfort food and watching sunsets in the hot tub and helping my body learn how to work hard and rest hard in equal measure. Thank you for being one of my most ardent supporters and biggest cheerleaders. I feel the force of your belief in me like the roots of a creosote. Thank you for loving me so much and for letting me witness your life and becoming.

Joon Seo, thank you for being the one I build a home with, and for your commitment to me, to us, to our lives. Thank you for all of your support, explicit and implicit, throughout this process. Thank you for teaching yourself to make gluten-free fried chicken and biscuits and gravy and tiramisu—I know that this is just one of the many ways you show your love to me, but it sure is a delicious one. Your support—mental, emotional, physical, and gastronomic—has fueled me and made it possible for me to put this into the world. I'm so grateful for you.

Zephyr! I can't even begin, dude. There are no words for the level of support, care, and love you've given me. Our connection in this world and in others is beyond anything I could have imagined, and I don't know where my life would be without you. I have more to say here than I can possibly put into printed words. Listen with your spirit, though; I know you can hear me.

Gabe—my best Judy. Thank you for being my Victorian sister, for listening to my multipart forty-plus-minute voice notes, for being such a steady friend to me. I will never forget the full-body relief I felt when we found one another again in the Bay, and ever since that point, you have been one of the fixtures of my life. Thank you for witnessing me, for believing in me, for supporting me, and for riding the waves of my wild life with me. Thank you for meeting my multitudes with your own.

Zel—I know that the transcestors brought us together for a reason, or for many reasons. I've learned so much from you, and your DNA is in this book as well, and not just where I've explicitly named you. I am so grateful to know you, I am so grateful for all that I have learned from you, and I am so grateful that we're alive at the same time. What a gift you are to this world, and to the transcestors, and to me.

Mars—thank you for your stubborn insistence on me, thank you for showing me that more and more and more is possible, thank you for pushing me to greater and greater visibility. Thank you for being willing to tell me what I can do better, and for being willing to hear me when I do the same for you. You're a joy and a light in my life, but also in the lives of so many others. I am so grateful we met.

My deepest gratitude to my editor, Gillian Hamel, who believed in me and my work enough to reach out, and who was such a helpful guide through the process of writing and publishing my first book. Having an editor who was aligned with me was crucial for this process. Thank you.

My gratitude to Cassandra Snow, Charlie Claire Burgess, and Gabriel Mac for your support and encouragement, particularly at the early draft stages of my proposal. Each of you encouraged me at critical junctures. This book would not be here today without your support and care.

I am so grateful to Sphere + Sundry and especially Kaitlin Coppock, who was willing to let me adapt her brilliant and effective bathing ritual for chapter 4. The magic of S+S is also deeply in the DNA of this book, and fueled many a writing session from the proposal all the way through the conclusion, and then the editing process as well.

My thanks, love, and care to all my clients. Every single person that I work with changes me on a deep, soul level. Each of you has informed this book, my work with each of you informs all of the rest of you. Thank you especially for every time you told me something *didn't* work for you. These moments help me become more nimble, more creative, more expansive in my offerings. Thank you.

Thank you to all the teachers who have held me through the years. To my doula mentors Erica Livingston and Laura Interlandi—you were the first to believe in me, you trained me and held me and saw me and anchored me. Both of you are written throughout this book, but also throughout my entire career. Thank you. I want to also thank Ylva Mara Radziszewski, Diana Rose, Sarah Faith Gottesdiener, Julia Inglis, Chauna Bryant, and Lindsay Mack for the many ways each of you has informed my practice.

There are so many more to thank and so much more gratitude to be offered. This is just a short sampling of the many human, nonhuman, and more-than-human beings who contributed in ways large and small to me, to this book, and to the process of our becoming. My gratitude is endless, but alas, my word count is not. Thank you to all those I have named and the many more I have not. I am so profoundly grateful to be alive in this time, to be writing, to be supported, to be able to witness the beauty and the struggle of these times. Thank you, thank you, thank you.

PART ONE

ritual basics

1

Introduction

If you are reading this, it was written for you. Regardless of how you identify, regardless of your spiritual practice or lack thereof, regardless of your relationship to or history with gender, this book has something in it for you. Even if you think that you are not trans enough, even if you think that you're not spiritual enough, even if you think that ritual is irrelevant to your experience, this book is for you because you are a human, your body is made of the stuff of this earth, you're connected to thousands of gender-expansive people across the entirety of human history, and because you, dear reader, have a relationship to gender.

It was March when we gathered on the beach, a group of twenty or so folks who were all coming to that place at that time to release names, pronouns, or outdated understandings of ourselves that no longer served us. I designed the ritual for that exact time and place—the beach is, after all, a *transitional* space where many elements come together, mingle, and separate again. We gathered stones along the shore, then I passed out chalk that folks could use to write on the stones whatever they wished to release. As a group we allowed ourselves to recollect all the beauty, all the good that we had experienced with that name, pronoun, or identity. Even if it no longer resonated, it had been with us through some significant parts of our lives, and that was worth honoring.

With that sense of gratitude and honor, we released those stones to the Pacific, knowing that the chalk would wash away in the waves. Newly unburdened, we each called in what we claimed instead—new names, new pronouns, new ways of being our current, whole, and authentic selves.

The folks who attended that day were not of any uniform type or identity. Some released their stones gently, allowing the tides to lap them away. Others hurled their rocks as hard as they could. Some danced along the shoreline, laughed, cried, hugged others. Regardless of why we were there and what we were releasing, each of us left that space changed. We were, each of us, changed by the power of ritual.

Gender rituals are a powerful form of medicine. In the past I've struggled to talk openly of rituals—their importance, their power, their potential—because I fear that I may sound hyperbolic. In my experience as a gender doula, I have designed and held many rituals for folks—sometimes writing the rituals for a client to do privately, sometimes officiating for a group or a community, and sometimes conducting the ritual in private for a single person, together in person or through the magic of the internet. Regardless of whether the ritual was on the beach, in a hotel room, in the desert, or on a video call, it was beautiful, intense, powerful beyond words.

In gender ritual, I have watched folks peel back layers of trauma. I have watched them find love or connection with their body after assuming those avenues were forever closed. I have watched folks resource themselves enough to enter the deep underworld journey of surgery, or bring their community around them to affirm their new name and pronouns. To hold ritual space is to constantly tread in the liminal spaces—the space between breaths, the space between life, death, and rebirth, the space between our bodies when we hold one another, the underworld.

But if that sounds like . . . well, a lot, please know that to *engage* in ritual and to *conduct* ritual on behalf of someone are two different skill sets. And my goal with this book is to provide you with rituals that are accessible and possible for you, no matter what you believe, no matter your level of experience. You're allowed (and welcomed) to take this entire book and the rituals within it at your own pace.

So, what's *your* relationship to ritual? Before I tell you how I define ritual, take a moment to think about what ritual means to you. How do you define it? What rituals do you do, and how often? Who taught you those rituals? What do they mean for you?

Humans are creatures of ritual. We tend toward them, in spite of ourselves, our reservations, our beliefs or lack thereof. Every culture has its own rituals, whether or not they are acknowledged as such. In modern Western contexts, we put a great deal of importance on certain rituals, but we don't acknowledge them as such. Our relationship with ritual is deeply impoverished, and this is reflected in our gender rituals.

Take, for example, the recently instantiated ritual of the gender reveal party, a ritual around announcing the supposed sex of a baby to one's community. These rituals took hold, at least in part, because they are ripe for social media–ready documentation. The dramatic photos of the cake being cut, the smoke bombs being set off, the plane releasing a plume of blue or pink smoke. Broadly derided by the trans community but cherished by a certain sect of expectant parents, these rituals have caused multiple deaths, wildfires, and other accidents across the country.

There are a few gendered coming-of-age rituals—sex-ed in health class, learning to use period products for the first time—but most of the rituals around adolescence depend on the community one happens to be born into. Unless you belong to a culture that performs something like a bar/bat mitzvah or a quinceañera, then adolescent gender rituals in the West are often far less rooted and more dependent on individual circumstances and family cultures.

In adulthood, this fractured sense of gender ritual persists; maybe, for some, a bachelor or bachelorette party. Maybe for others a coming out, a first pride parade, starting hormone replacement therapy (HRT), or pursuing surgery. For some, daily routines such as shaving or applying makeup may feel like ritual or gender, and for others, those things may not even register as a part of life. Maybe nothing I've said so far resonates with you, your life, your individual circumstances. This is what I mean when I say our relationship to gender rituals is deeply impoverished.

And yet gender impacts so much of our lives. Whether we broadly identify with our assigned gender or not, whether we perform our gender in ways that are expected or un-, many studies across academic disciplines have revealed that gender impacts us throughout our lives in myriad ways. Your *perceived* gender identity and performance constrains the types of jobs that you get, your pay at those jobs, how you're treated by strangers, and also how you're treated by supervisors, friends, and coworkers. Gender, both the performance and the identity, can foreclose certain possibilities for us and open our worlds to others. It seems strange that something so impactful across personal, professional, emotional, mental, and spiritual realms would be largely ignored in ritual.

What Is Ritual?

There are many ways to define ritual, but my definition is this: some intentional set of actions done for a specific purpose. This is a wildly broad definition, but I find its broadness to be helpful, because ritual can be almost anything. A ritual can take place wherever it needs to, it can last as long as it needs to, and it can be either planned or intuitive.

A ritual can be done alone or with another person or with an entire community.

What separates ritual from any other action is the intention and the purpose. I can light a candle because I want the room to smell nice, or because I want the ambiance of the flickering light—neither of these is necessarily ritual. But if I light a candle with a specific intention to support a carefully chosen purpose? That is, or could be, a ritual in itself, or an action taken in service of a larger ritual. It is the intention and purposefulness with which the act is performed that transforms it from the mundane to the sacred.

We all engage with ritual in one way or another. Perhaps you have an annual ritual for your birthday, or for a holiday such as Christmas or summer solstice. Perhaps your rituals are brushing your teeth at night or checking in with your mom once a week. Or perhaps you engage in more organized rituals as part of your cultural, religious, or spiritual practices.

Regardless, ritual allows us to mark and track time. It allows us to locate ourselves in time as well, to see our growth, to connect to ourselves, our world, our community. Sometimes, ritual creates a container within which we can do our deepest work, heal trauma, shift ancestral wounds. With ritual, you can connect your whole self—mind, body, spirit—to the earth, to community, or to a higher power. Ritual often transcends human differences and allows us to connect with some deeper aspect of ourselves, and therefore, some deeper aspect of one another.

The power of ritual has made it a target for destruction during colonization and the creation of a white supremacist state. Ritual is a threat to the colonizer and the oppressor, and therefore existing ritual must be either destroyed or assimilated into the culture of the colonizer. Across the colonization process of many lands and peoples, when assimilation failed, colonizers simply did their damnedest to destroy the existing cultural celebrations and rites.

Ritual is a powerful aspect of culture. If it weren't, it wouldn't need to be a target in the process of colonization and genocide. Even when attempts are made to eradicate the cultural and spiritual rituals of a group, versions of those rituals will often live on. Unless you manage to completely annihilate an entire group of people, ritual will find ways to continue. When a group of people is under attack, their rituals become even more vital, even more treasured. After all, a group of people without rituals are just a collection of humans. Ritual binds us together.

The Lack of Rituals in Modern Western Culture

I was born in Tulsa, Oklahoma, and raised in the southwest corner of Missouri. My paternal side came to America from Ireland four generations before I was born. My maternal side came to America via Canada and, before that, England, in the nineteenth century. Which is all to say that my personal context is very much white American. My family was conservative Christian, and thus I grew up with a set of rituals that did not resonate with me. I began questioning Christianity when I was a young child. While I enjoyed certain aspects of evangelical ritual, the whole enterprise didn't suit me. By the time I was ten years old I was just going through the motions to appease my family.

Yet as an agnostic teen and later an ardent atheist, I felt the lack of ritual in my culture. The culturally acceptable rituals that existed all seemed to be organized around spending money. Although Jesus had been the reason for the holiday season in my family, several years of working retail around the holidays made it quite clear who the season was really for. In American culture, at least in Missouri, the most exciting ritual was getting your driver's license, then becoming old enough to legally drink. After age twenty-one, what was there? Maybe marriage, if you waited that long. Maybe baby showers. Maybe, depending on your denomination, church rituals. Maybe graduation from college.

Moving to Los Angeles taught me quite a lot about ritual. My first job was as a live-in cook for a Jewish family in the San Fernando Valley. I learned how to cook for Passover and prepare for Shabbat as I participated in the rituals of a completely different culture. I was still an atheist at the time, and I met quite a few Jewish atheists. "I'm culturally Jewish," they would tell me. "But I don't believe in god." This boggled my mind, as a person who was raised evangelical. I never knew that cultural rituals could be separated from their religious background, or that one could participate as a form of social belonging instead of worship. Witnessing it changed something in me, because for the first time, I could separate spiritual ritual from organized religion.

A few years later, when I felt spirituality calling to me, I was lost. I wasn't going to commit to an organized religion—I had way too much religious trauma for that. While a lot of what resonated with me landed in the realm of pagan belief systems, most of them were blatantly stealing ideas from Indigenous or Asian cultures, or were thinly veiled versions of white supremacy, or were far too organized for my comfort. Eventually, in my desperation to root into something that didn't compromise my values or activate my trauma, I asked myself, "What do you know to be true?" The answer that I eventually arrived on was transness and queerness.

Thus began my ongoing journey of creating my own spiritual practice, rooted in what I call the sacred knowledges of transness. Some of the rituals of this spirituality are very common—cutting or growing your hair into something that feels daring and unexpected, taking an HRT shot or pill, purging your wardrobe of clothes that no longer resonate with you, sacrificing parts of your body on the altar of your authentic embodiment, taking a new name, new pronouns—but are not always considered rituals.

They should be. As I moved through my own healing path, I received my calling to become a gender doula. In this role, I have heard countless folks say, in one way or another, "This feels so momentous, but the people around me are treating it like it's no big deal. I know they want to be supportive, and I don't want people to question me or be mean about it, but it is a big deal. I want someone to treat it like a big deal." This, to me, represents the hunger for ritual. Having a loved one act like you don't understand the gravity of gender exploration or transition can be deeply harmful and painful. But when everyone around you more or less just says, "You're transitioning? That's fine," it can leave you feeling empty.

The fact is, we're all hungry for ritual. Without ritual we just slide from one part of life to the next. Without ritual it's easy to lose the thread on what is important, what matters, that we matter, that our paths matter. Ritual isn't about making a big deal over every little thing (though, honestly, almost every aspect of transition and transness are big deals). Rather, it's about honoring the importance of something in our lives. Through ritual, transition can be honored and celebrated for the sacred process that it is in a way that is life-affirming, identity-affirming, and crucially, community-affirming.

In the years since I began my own spiritual journey, moving along the path of transness as a site of sacred knowledge, I have come back to ritual again and again. I have developed rituals to hold myself through particularly potent portals, and I have also used the skills I've developed to hold others. Through ritual, I have helped folks move through the fear and anxiety that so often comes before a surgery, no matter how hard they fought to obtain it. Through ritual, I have helped folks make their bodies into their allies for the journey of HRT. I have helped folks and myself reclaim sexuality, shed old identities and welcome in new ones, integrate new experiences, process out old traumas.

Ritual is potent, in my opinion, regardless of your belief system. Ritual is a technology available to all humans and to other animals as well. Famously, elephants, dolphins, crows, magpies, and others display ritual behaviors. There are likely countless rituals in the animal world that humans simply don't have the knowledge or the tools to properly observe. Ritual is ancient and primal, something that we return to again and again and that we see evidence for across time and space, wherever life exists.

There are a few common objections to performing ritual. One is, "What if I'm not spiritual?" You may have guessed this already, but there's no need to be spiritual to engage with ritual. Much like the folks I met who were culturally but not spiritually Jewish, ritual is available to all. Singing "Happy Birthday" can be a ritual, as can preparing for a date by washing your body, fixing your hair, and applying products to your skin. Even if you are not spiritual, I guarantee you already engage in ritual. And each of the rituals in this book can be customized in whatever way makes sense to you.

Another common concern is that doing ritual will feel weird or awkward. It may indeed! Even some of our most culturally accepted rituals, like marriage ceremonies, can feel weird and awkward. Feeling weird or awkward about a ritual will not make it less effective, and it does not indicate that you're doing it wrong. Things feel normal and natural when we have experience doing them. If doing ritual feels weird, and you want it to feel natural, you probably just need to do more ritual!

Group Ritual Versus Private Ritual

When we gathered together on the beach to release the names, pronouns, and identities that no longer served us, part of the magic was in the doing of the thing *together*. Coming together with other humans, breathing together, singing together, dancing together—these actions help us recognize that we are not singular beings on isolated paths, but parts of a living whole, parts of a web of energy and connection. Doing ritual with other people allows us to access a specific state of mind where we can maintain our own specificity while also recognizing the depth and breadth of our connection.

Yet solitary ritual can be powerful as well. The quiet of a ritual without other humans present can help you to be clear on, and in, your own energy. It can allow you to cultivate the courage that you need to move through a threshold moment, like your first HRT shot or calling in a new name for the first time. Solitary ritual can help you develop and tend to your own unique connection to the earth, to your spiritual root, to your ancestors, or to your trans ancestors (transcestors; see chapter 12).

It is not that ritual in community is better and ritual in solitude is worse. Rather, in the hyperindividualistic culture of the Western world, built on the foundation of white supremacy, it can be easy to default to solitary ritual and miss the benefits of group ritual. I encourage you to actively avoid this. While many of us struggle with community—finding it, maintaining it, and managing conflict within it—humans evolved to be communal creatures. Rituals as simple and timeless as sharing a meal benefit from human connection.

Yet ritual, especially gender ritual, with other humans may not be available to you for a variety of reasons. Many of the rituals that we engage in around gender exploration and transition are intensely private, and we may not feel safe sharing that level of depth within community. For some of us the people we would choose to witness us may live far away. For others the logistics of public ritual may be prohibitive. In this moment in time, for many folks, a public ritual may not feel physically safe due to antitrans violence. And finally, we may not have people in our lives that we trust to hold the fullness of our identities.

For the purpose of accessibility, I have made most of these rituals amenable to solitude, and have kept rituals that assume community involvement to a minimum. But that doesn't mean that you can't or shouldn't use these rituals in community. We must all seek the balance between the vulnerability and challenge of community space and the safety and predictability of solitary practice. Both are good, and the exact balance will depend on you and your needs. There's nothing wrong with keeping the most vulnerable version of a ritual to yourself and creating a less vulnerable version to be done in community.

What This Book Contains and How to Use It

This book is meant to contain everything that you need to do gender rituals. This means that if you are reading it, and you've never done a single ritual before, or at least have never done one on purpose, this book should have all the basic knowledge and steps you need to perform gender rituals. If you're a seasoned ritualist, there may be sections that are review for you. Still, I recommend at least skimming the first part of the book, which covers when to use rituals, how to customize them, and some foundational skills that will be helpful in your journey with ritual work.

In chapter 2, I attend to the temporal aspects of choosing when to do ritual, how often to engage in a specific ritual, and how to create boundaried containers of time to support consistency in your ritual practice. Frequently, folks believe that a ritual is a sort of big one-and-done ceremony. While this is certainly one type of ritual, and the best type of ritual for specific applications, it is not the best form for all needs. When we need to shift a deeply ingrained pattern, smaller, more doable, more consistent rituals can be key.

Chapter 3 covers how to customize a ritual. This is an important chapter because even though I have worked very hard on the rituals included in this volume, there may be elements of them that do not resonate with you. This section gives all the information you need to compose your own rituals, to write them out and customize them with ritual elements that resonate with you. Crucially, this section discusses cultural appropriation and choosing ritual elements that are appropriate to your own background and resonances.

This section also covers making ritual doable for you, specifically, and how to leave behind that which doesn't work for you. We will cover choosing a proper venue for a ritual, when to do private ritual versus community ritual, and when to involve an officiant or space-holder in a ritual. I provide questions that will hopefully create the space for you to discern for yourself how you want to customize the ritual.

In chapters 4 and 5, we get into the foundational skills needed to engage with the rituals in this book. This section is separated into two chapters because some of these foundational skills are best used daily, or at least regularly, while the rest are more centered around occasional rituals. In chapter 4, we cover cleansing, grounding, and shielding. I offer options for engaging with these skills in various ways, and ultimately suggest working with these skills for daily energetic hygiene, then doing larger or more intricate versions of these skills when desired.

In chapter 5, we cover opening and closing a space, working with intuition, building and tending an altar, creating sigils and petitions, working with tarot and oracle cards, and journaling or otherwise recording rituals or other intense experiences. Even if you have never created an altar or pulled a card, this section will help you find your way into whatever practices will be supportive of the rituals you wish to perform.

The second part of the book covers many specific rituals that are related to gender. You may see uses for some or many of these rituals that are aside from or in addition to gender; that's great! Once you get the hang of it, you can make your own rituals for almost anything you want. I have tried to offer a broad variety of rituals that cover many of the aspects of the gender exploration and transition experience, from attuning to your own energy (a vital foundational skill for anyone of any identity, belief, or experience) to preparing for surgery. Take a look at the table of contents to see all the different names of the rituals. I've tried to make them as descriptive as possible.

For the most part you do not need to do these rituals in any particular order. As mentioned, attuning to your own energy is an important skill and for most folks is wise to do earlier than later. But if you're reading this book days before surgery, I can imagine you may want to start from the presurgical rituals, and there's nothing wrong with that.

Some of these rituals, for example daily HRT rituals, are written to be quite specific to medical transition, while others are related to other parts of gender exploration and transition, like the ritual for releasing your attachment to old stories about yourself. If you feel drawn to any ritual, I recommend you read through it once or twice, then gather the materials and set up the circumstances to make it happen. And if you never do a single gender ritual, but you start a regular practice for cleansing, grounding, and shielding as outlined in chapter 4, then reading this book will have been worth it.

A Note on Terminology

When discussing both gender and spirituality, there are many linguistic pitfalls. For example, the word *trans* can be understood to include or exclude many different folks and experiences. To be clear, I use *trans* as a shorthand for gender expansive, nonbinary, and trans people. Language around gender shifts rapidly, and what is cutting-edge will often be dull and outdated mere months later. So if you read this in a future time when my terminology seems dull and outdated, I hope that you will view these words favorably. My intention and goal, always, is to be as inclusive as possible without losing the meaning of my ideas.

Similarly, the word *spiritual* can be quite tricky, especially for a community with so much religious trauma. Further, in the US context, alternative spiritualities are often rife with white supremacy, capitalism, and other forms of oppression. Words like *blessings* or *prayer* are associated by some with organized religion and by others with New Age nonsense.

To be very clear: To me, all beings have access to the spiritual, because the spiritual is the web of connection among all beings. My own belief systems are rooted in transness as a portal to an animistic understanding of the world. All beings—human, nonhuman, more-than-human—have animacy, consciousness, connection. People feel that connection and experience it in different ways, but we all have some access to it. That access point, that potential for rooting, that connection to the web of all things—that is what I refer to when I refer to the spiritual.

I do not want or need anyone to believe what I believe. There is no "book of answers" that I turn to, no ultimate authority, no need to proselytize. Rather, I have rooted into the connections that exist in my sphere, and built my relationship to the mystery from that place, such that it is only always my own. If you're wondering how one does such a thing, I have provided a ritual for attuning to your own energy. That is an excellent place to start.

You do not need to believe as I believe to use this book. You do not need to identify as I identify. Take the pieces of it that speak to you. Iterate, keep notes, adjust. Over time, you will build your own relationship to ritual, to yourself, to your web of connections. That is where you will find your power.

Conclusion

You may have noticed that the positioning of this book is extremely trans. You may think that means that the book is only useful for or by trans people. But I started this introduction by saying that if you're reading the book, it's meant for you.

There are a few reasons I started there. First of all, almost everyone who has ever been trans had to, at some point, realize that about themselves. For most folks, at least those around my age, that realization wasn't particularly easy to come by. So if you're a person who doesn't identify as trans, well, that could shift. I wanted this book to be available for you as well. Further, if that doesn't shift and you explore your gender and land on the deep knowledge that you are, indeed, a cisgender person, that's actually great too. I think a lot of the violence that trans people experience could be mitigated if cis people took the time to explore their own genders and do whatever it was that they needed to in order to live authentically and well.

I believe that anyone who picks up this book will be able to find some support within and through it. So I reiterate that if you are reading this, it was written for you. If you are reading this, there is something, or perhaps many things, tucked into this text that will serve you, even if you don't yet know why. The journey is half the fun.

Ritual has been an absolutely profound part of my journey and the journeys of so many of my clients, students, and community members. My intention for this book is that it will bring the magic of ritual to many more folks, and that it will allow you, dear reader, to experience this magic for yourself. My intention for you is that you will receive from this text the exact thing that you need from it.

Happy ritual-crafting!

2

Timing Your Rituals

Introduction

When you think about ritual, how often do you feel that it should be a part of your life? There is no particularly correct answer—each of us has our own relationship to ritual—but it is easy to consider only the most dramatic rituals and forget that in reality, life is made of endless small rituals that direct our time and attention. There's value in recognizing that even rituals that take very little time or energy will permanently alter our lives if we engage with them consistently enough, and that big or dramatic rituals are not necessarily the best for each occasion.

In my experience with holding ritual for and with clients, this is the piece I have most often found that folks are surprised by: that *duration* and *consistency* are often a huge part of the impact of ritual. Instead of thinking of ritual only as something that you do once you have reached a certain milestone, consider it to be part of how you learn, grow, and get to know yourself. Ritual can be one tool of many that you use to help tap into your own knowing, into the wisdom of your transcestors, into the flow of your own desire and energy. (For more about transcestors, see chapter 12.)

If you spend an entire lunation (new moon to new moon, full moon to full moon, etc.) intentionally connecting to your body daily, what might shift? At the very least, you will likely be connecting to your body more consistently than you were before, and in that connection, all kinds of new information can emerge. That new information will in turn shift how you show up in the world, opening up new possibilities that can necessitate new rituals. In this way, we loop ritual through our lives, using it as one robust toolkit for personal growth, change, and celebration.

This chapter delves into the relationship between ritual and timing across multiple dimensions of consideration. The purpose is not to give rigid instructions about timing, frequency, and duration of ritual, but rather to guide your decision-making around timing. We discuss when to use ritual, how to make decisions about length and duration, how to time a ritual, and how to weave ritual in as a consistent element of your life. Working with these different elements of ritual can help you create a deep, effective practice that works for who you are right now.

When to Use Ritual

I don't believe there is a "wrong" or a "right" time to use ritual, so the quick answer is anytime it serves you, your goals, your life, your community, or the earth. But that might feel too broad to be workable, so let's get specific.

The first element to consider is why you want to engage in ritual. For example, do you want to celebrate something? Mark it? Grieve it? Do you want to create a new habit in your life? Deepen a practice? Connect to yourself or to another being? Something else entirely? Ritual can be used for any of these reasons, so getting clear on the why can be really useful in making logistical choices about timing. If your *why* doesn't feel clear, take some time to journal, process with a friend, or meditate. The goal isn't to have a perfect *why* but to understand enough about your intention to make initial choices about the details.

In general, if you want to mark a specific experience, event, or turning point in your life, a one-time ritual might be the way to go. These rituals often carry the signature of a distinct before and after. However, if you want to shift some internal pattern, create a new habit, or engage in deep change work, a smaller and more sustained ritual over a specific container of time will be more effective. These types of rituals support more incremental change: small steps that add up to significant results.

For example, a ritual to release a name, pronoun, or identity label and claim a new one could be a one-time ritual, whether done in community or not. This ritual marks a specific experience or event, and it has a distinct before and after. It doesn't mean that you could never change your name, pronoun, or identity label again, but it is like drawing a distinct line in the sand of your life.

By comparison, a ritual to develop a better connection to your body would probably benefit from consistency and repetition. A smaller ritual done regularly for a period of time helps your entire self reorient to a new energy or habit. Further, if you're trying to bridge a very large gap between your current reality and your desired reality, smaller steps can take what feels impossible and ground it into reality. Using the body connection example, if you're utterly dissociated, coming back into your body fully and suddenly is probably impossible and definitely inadvisable. Coming back a bit at a time, starting with the parts that feel safe to inhabit, will make for a more sustainable and durable change. The ritual timing and duration can both reflect and support this.

These aren't rules that are set in stone. Using the above examples, a smaller, private, consistent ritual for preparing your body and mind to release one name or pronoun may appeal to you more. You could then cap that period off with a larger ritual done in community. Or you may bookend a period of doing small daily ritual with a larger opening and closing ceremony. There are so many different rhythms available, and the only requirement for the type of ritual I'm outlining here is that it works for you, specifically.

Loosely speaking, I use ritual when:

- I need to change or shift something in my life.
- I need to be consistent in a practice.
- I want to create a devotional relationship.
- I need to mark a specific moment, change, or experience.
- As a part of my spiritual practice.

The more you use ritual, the more you will hone your intuition about when it is the specific tool that you need. Again, start with whatever rituals in this book speak to you, and use that as a starting point. Pay attention to what works, what feels right to you, and what doesn't. You can use these details to tailor your ritual practice to your own unique specifications.

Ritual Length and Frequency

To be completely clear—the length of a ritual does not determine its depth or efficacy. In fact, some of the most effective rituals I've ever done were only a few minutes at most. Depth and efficacy can be enhanced in all sorts of ways. One of my favorites is consistency.

Using a daily or regular ritual during a specific period of time can effectively shift your energy or your path, even if the ritual seems small to you. To continue the example from the beginning of the chapter, you might decide that from one new moon to the next (for one new moon lunation), you will spend five minutes each morning connecting with your body. The ritual around this may be simple. Sitting in a specific place, listening to a specific song or playlist, moving in intuitive ways or going through a prescribed set of stretches, lighting a specific candle, applying a specific perfume or scent—all of these can be ritual elements that mark this practice or set it apart from the rest of the hustle and bustle of your day. When you make the time and effort to show up consistently over an extended period of time, even the seemingly smallest changes can have tremendous results. This too is the magic and the power of ritual.

I know that my neurodivergent friends are often intimidated or excluded in conversations of consistency, so if that is you, read on. When I say "consistency," I don't mean that you have to do the ritual every single day or else you will be a failure. I mean doing the ritual on a schedule that is consistent by your own standards. For some, this will be every day. For others it will be a few times a week on whatever days you have the capacity for it. This is purely subjective, and you will not benefit from beating yourself up about not being adequately consistent.

There are a variety of ways to approach consistency. You can attach your ritual to a daily habit, or to a part of your schedule that tends to be consistent. I often recommend doing something that you can do while brushing your teeth, or in the first moments when you wake up or go to bed. This type of habit stacking can make it easier to develop consistency with a new practice.

It's also helpful to create a specific container of time for a particular ritual purpose. Indefinite consistency may be impossible, but a boundaried period of ritual can create more possibility. Additionally, although repetition can lend potency to a ritual, it can also make the process feel rote. Choosing a specific time container for a practice gives you a clear beginning and end date, which creates space for you to check in with yourself and change your mind about whether or not you want to continue the practice. A time container creates a natural pause where you can decide whether or not the ritual is complete, if it needs to transform, or if it needs to continue.

As I mentioned, my favorite time container is a lunation—new moon to new moon, full moon to full moon, or some other moon-based timing.

There are many reasons for this. The moon changes consistently, and an entire moon cycle is about twenty-nine days. Humans have timed our lives by the moon since before recorded history. Though many of us currently use the Gregorian calendar (one year is January 1 to December 31), this is a recent development in the long span of human history. The moon was here before that convention and has guided countless human beings through time. The moon influences the waters of the earth, the moods and rhythms of the life on this planet, and our consciousness itself. If you're curious about the influence of the moon on human psyche, I highly recommend the episode "Your Consciousness Comes From the Moon" from *The Emerald* podcast.

If that doesn't speak to you, consider that there are many things you can do consistently for twenty-nine days that may be hard to do consistently for longer than that. The moon is a helpful way to mark this period of time. And hey, if you work with the moon a few times in this way, you may find yourself more drawn to, or interested in, the more mystical things I am referencing here. If you're interested in working with the moon, I highly recommend *The Moon Book: Lunar Magic to Change Your Life* by Sarah Faith Gottesdiener. Sarah is a friend and teacher of mine, and her methods for doing lunar work are absolutely foundational to my practice.

Part of maintaining a consistent ritual practice is being aware of your personal capacity. Many of us tend to be out of touch with our bodies, energies, and capacities, and therefore over- or underestimate what we actually have space for in our lives. To compensate for this, I recommend choosing something that seems doable, and then making it even smaller or more doable. Why? Most of us operate to some degree in the realm of *should*. As in, "I should be able to spend thirty minutes in ritual every morning, so I will just get up thirty minutes earlier so I can do this." In my experience, this is a recipe for shame and failure.

Instead, ask yourself what is the easily doable amount per day? For example, does it seem easy to do ten minutes of ritual per day? How about five minutes? What about ninety seconds? I often encourage folks to choose whatever feels laughably simple: so simple that it would almost be silly not to do it. Start there and do it consistently for a moon cycle, or a week, or a month, or two months. Keep track of what changes during that time, then reassess.

If you feel organically like doing more, then do so. If you miss a day, come back to it the next day. It's not a big deal; don't beat yourself up. Just consistently orient yourself back to your ritual again and again. If it is not working for you, change it! Make it shorter or longer. Try again. Adjust as necessary. If you want to, take a break. Note what changes, and note what doesn't. This type of reflection can help you build a practice that supports you, specifically.

How to Time One-Off Rituals

When you've ascertained that a ritual would be helpful, and you've decided the duration of the ritual, you are ready to choose the exact timing. For one of the larger one-and-done type of rituals, you might consider whether or not you want that ritual to occur on a day that already holds significance to you. For example, a name and pronoun release ritual might be a great way to spend your birthday. Other rituals will have their timing impacted by factors outside your control. If you're doing presurgical ritual, I recommend timing it close to the surgery, but not so close that you'll be rushing to get it done before you go to bed the night before.

Some folks may wish to time their rituals to specific astrological events. The lunation timing I mentioned before is one version of this. Another option is important solar days, like the spring or autumn equinoxes or the summer or winter solstices. You could time your ritual around specific zodiacal seasons, like doing a release ceremony during Scorpio season or a series of self-love rituals during Taurus season, or around the way that specific parts of the cycle of the year impact the plants, animals, and weather in the place where you live. Perhaps a spring ritual for a new beginning, a deep winter ritual for a releasing, a summer ritual for celebration, an autumn ritual for mourning and underworld work.

There's also a form of astrology called electional astrology. Astrologers trained in this skill set can look at the movement of the planets and stars during a specific period and advise you on the best timing for specific purposes. This type of timing work takes a lot of expertise and practice, and it doesn't come cheap. If you want to get an election for a specific important ritual, I recommend choosing your astrologer carefully, as elections are a complex and specific art that requires specialized training. If you want to support a reliable trans astrologer, I recommend Ursula Rising (www.ursularising.com).

When considering the timing of a one-time ritual, it is helpful to determine how much time the ritual will take and how resourced you are or will be. You probably don't want to do a four-hour ritual at midnight if you're a certified morning person, right? I will often start from planning whether or not a ritual feels more appropriate for daytime or nighttime, then I will see where I can feasibly do that ritual in the general rhythm of my days or weeks. A longer ritual needs more focused attention on the specificity of the timing, but in general, some helpful considerations include:

- What time and place will I have privacy? For how long?
- Does this ritual need to take place somewhere other than my home? If so, do I need to make specific arrangements based on hours, weather, crowds, and other factors?
- Do I need access to any particular resources or amenities during this ritual?
- Is it crucial that this ritual be private or uninterrupted?
- Is there a physical geographical location that feels meaningful for the purposes of this ritual?

If a ritual can be done at home, and you can attain privacy at home, that can solve a lot of issues. But if a ritual needs to be done in a public place, or in a place that lacks some of the basic amenities we often take for granted, some additional planning may be needed. Sitting with the questions above should help you decipher the specific considerations for your ritual.

As someone who uses astrology and works with planets, I will often key my rituals to planetary days and hours for additional specificity. A ritual that deals in love, pleasure, beauty, abundance, justice, or balance would be a Friday daytime or Monday night affair for me, as those are the times that Venus rules. A ritual for boundaries, goals, and hard work could be on Saturday daytime or Tuesday nighttime, the times of Saturn. Additionally, each day is divided into planetary hours—blocks of time that are ruled by specific planets. You can look up the planetary day and hour at www.planetaryhours.net if you would like to use this simple but powerful timing technique.

How to Time Ongoing Ritual

Timing for an ongoing, consistent, or daily ritual requires a different set of considerations. It is helpful to be aware of your body, capacity, and schedule to make this type of ritual work as intended. For my own practice, I've created a series of morning rituals that are vital to my work as a gender doula. These rituals are elastic and shift with my needs, desires, and intentions. If I want to do ritual toward a longer-term goal, adding it to this time can be a helpful and simple way to maintain consistency.

When timing your consistent rituals, consider what feels doable to your body. Some folks do well with guided meditation or other guided ritual, while others do best when they can guide themselves. Some folks need to move their body in order to accomplish anything, while others do best sitting still. Some folks need to feel themselves doing something material, while others can swim in the realm of the mind or emotions. None of these are incorrect, none of them less powerful than any other option. What matters above all is that you find what feels doable, possible, and easeful—start there.

Next, consider the rhythm of your energy and timing. What time of day do you find yourself to be most resourced? Are there parts of the week that you can anticipate you will be exhausted or scattered? My weekday evenings, for example, include a lot less ritual than my weekday mornings. Weekends are more of a free for all—my energy is more elastic, so my schedule can be as well. Over time, you may find that a more exhausted energy works better for certain types of rituals, and that a sharp alert energy works best for others. You may need to engage in some trial and error to discern your own patterns, and those patterns will most likely shift over time.

Take some time to notice what rituals you are already consistent with. None of these rituals have to look particularly special, spiritual, or even intentional. Perhaps you sit down to eat breakfast in the morning most days. At one point I noticed that when I sat down to eat breakfast, I would spend the time reading my emails and scrolling on my phone, and that this was distracting and often distressing. I chose to shift my phone-scrolling time to book-reading time, and suddenly I was getting through all these incredible books that I had purchased and never gotten around to starting. Later, I decided that I wanted to journal instead of reading, and I managed to keep a consistent journal for the first time in my life.

I offer these examples to show how something that is functionally a ritual, like scrolling your phone while you eat breakfast, can just seem like a default option, almost a nonaction. But when you examine it through the lens of ritual, these default options actually have a significant impact on your time, energy, and experience. Shifting at least one of these default moments during the day can give you a foundation for your ritual practice. If you understand the rituals you currently engage in, intentionally or not, you will develop a better sense of where your energy is going, and how you might be able to shift it. As you experiment with different rituals and options, you can change the energetic patterns of your days. Over time, this will change your life.

This can be woven together with our lunation example. Perhaps you too have noticed that you spend your evening relaxation time in front of a screen, and that after doing so, you don't feel particularly good. Let's say that you want the option to do screen-related activities, but you also want to add in another activity for at least some part of your evenings. But when you try to do so, you find yourself feeling resistant and lethargic. This is a great time to use a lunation rhythm. You don't have to give up your comforting screen time—simply decide that for the next lunation, you're going to try something else.

If you've not worked with lunation in this way before, it can be quite intuitive, but doing anything consistently for twenty-nine-ish days might feel daunting. When leading lunation-based rituals, I am often asked how one maintains the pacing and momentum through the entire twenty-nine-day cycle. One advantage of lunation work is that the ebbs and flows are naturally built in via the patterns of the moon. Your individual relationship to the moon phases may differ from conventional wisdom or what you see on the internet. For example, you might feel low energy on the full moon, and high energy on the waning moon. If you want to go deep on your own relationship to moon cycles, I recommend Sarah Faith Gottesdiener's *The Moon Book*.

But if you're not sure which moon phase to choose, you can simply go with a new moon for seeds you want to plant, and a full moon for harvesting or releasing. You will start with the moon in whatever phase you've chosen, then do ritual consistently for the following twenty-nine days until the next time the moon is in the same condition. To keep momentum for the full twenty-nine days, I find it to be most effective when I mix rituals of different type, duration, and intensity based on the moon cycle. I've included the following example for clarity.

EXAMPLE
New Moon Lunation Ritual Rhythm

Day 1: First new moon. Big opening ritual, cleansing and preparing, setting intentions, lighting candles, altar building, divination, and anything else that feels relevant.

Days 2–6: Daily smaller rituals. Adding a cup of water to the altar, saying a prayer or incantation, meditation.

Day 7 (ish): First quarter moon. Ritual to reaffirm intentions, adjust anything that has shifted, possibly more offerings on the altar.

Days 8–13: Daily smaller rituals continue.

Day 14 (ish): Full moon. Here I would do a robust ritual of some sort, often working with gratitude, abundance, or "fullness" in its many forms.

Days 15–20: Daily smaller rituals continue.

Day 21 (ish): Last quarter moon. Ritual for reflecting on the experience so far, shifting anything that needs to be shifted, gratitude, reaffirming intentions to finish strong.

Days 22–28: Daily smaller rituals continue.

Day 29 (ish): Second new moon. Close the ritual with gratitude and reflection.

Remember this is just an example. Try it and then adjust it to the rhythms that work best for you specifically. I hope that seeing it written out in this way will help you to get a feel for how effective this sort of timing can be, and how you might integrate this into your own life and practice.

Ritual As a Consistent Part of Life

So, when to perform rituals? Any time you want, and probably more often than you think you should. I have rituals that I perform daily or almost daily, others that happen weekly, moon-thly, seasonally—each ritual can happen on its own schedule and time frame. Not all of these rituals are related to gender though. You may have gender rituals that you perform once, and others that you perform consistently for some time. A ritual may feel deeply necessary for a period of life, and then one day you may find that you're complete with that specific practice, but ready for something else. For example, at one point I needed to perform sustained regular cord-cutting rituals for harmful relationship patterns that were deeply imbedded in my body. A full lunation of cord-cutting, following a similar pattern to the example above, cleared the pattern from me. I didn't need to continue with that same ritual and shifted to a more general form of daily cleansing.

This sort of natural migration from one set of rituals to another is nothing to worry about. It is normal and good that our rituals shift, change, and grow as we do. Many of my clients begin their hormone journeys with weekly rituals. For most, these rituals naturally shrink or drop away as the client becomes more comfortable with their HRT journey. For others, the ritual remains a cherished anchoring point in their week. What matters with any ritual is that it is still working for you specifically.

In this book you will find a whole host of rituals that you can use for specific needs. However, my not-so-secret plan is that, in doing some of these rituals, you will begin to get a feeling for what a ritual entails, and even for what your specific needs and preferences around ritual include. My hope is that you will be able to create your own rituals, customized to your individual preferences and needs, and feel the magic of a more ritual-focused life.

Your relationship to ritual will be as unique as your relationship to gender, and that includes the timing, frequency, and consistency of your rituals. The point of engaging in any ritual, including those in this book, is not perfection. Instead, the point is in the doing of ritual itself. Similar to the adage about journeys and destinations, it is your practice of ritual that will make an appreciable difference for you, not dogma about how long or how consistent a ritual should be. Your goal with the length and the consistency of rituals is, "What will I actually do? What will serve the function it needs to for me?"

A consistent ritual that is two minutes per day will generally have a greater impact on changing your life than a one-time ritual, even if it is dramatic and expertly crafted. But a one-time longer ritual may be exactly what you need to mark a shift or change in your life. Ultimately, it is up to you to discern what type of ritual you need, whether or not it should be consistently done over time or a one-off, and what elements it should entail. The following chapters will continue to help you define these considerations and boost your skills to make the necessary discernments.

3

Customizing Ritual

One of the aspects of ritual that trips a lot of folks up is specificity. It can feel like everything has to be exactly right, formulated in an exact manner, and that any deviation means that the ritual won't work. While specificity matters in certain rituals and cases, the type of ritual we're covering in this book is highly customizable. And that's not an accident!

Being able to create your own rituals or iterate on rituals made by someone else can be wildly empowering. It can create space for you to engage with your intuitive knowing or to weave in elements that you have a personal relationship to. When I create customized ritual with folks, we have a chance to talk about their preferences and what speaks to them specifically. While I have a general structure I work with for rituals, custom rituals are written fresh for each person. I will work with folks until the language is refined according to their needs and preferences.

Yet my years of work as a gender doula, teacher, tarot reader, ritualist, and facilitator have all taught me that there is no instruction that works for everyone. Further, each of us has our own relationship to the web of all things—I have facilitated for people who considered themselves spiritual but not religious, folks who described themselves as "scientifically minded; I don't really buy all this spiritual stuff," and for folks who were themselves ordained or otherwise recognized as a leader by an organized religion. I've been able to tailor ritual in all these contexts, in part because my mind-set when approaching this type of ritual is elastic.

In this chapter, I will lay out some of the contours of that elastic mind-set. When tailoring ritual, it is helpful to consider the general shape of ritual, what is and is not optional, what to avoid, and how to choose various elements of ritual as well as choosing a venue, deciding between community and individual ritual, and when to involve an officiant or other space holder.

Ritual Basics

If a ritual can be customized so thoroughly, does that make it less effective? Does it mean the ritual was meaningless to begin with? It is true that there are certain rituals that require specificity. Ceremonial magic, culturally specific rituals, and rituals that rely on particular deities or astrological elections generally require a lot of specificity and attention to detail. The rituals in this book, however, are rituals that you as an individual are using to mark your own life and your own process. The requirement for specificity isn't coming from any outside system, spirit, or deity. Instead, it's about being highly tuned in to your own self and doing what works for you as a unique being. In other words, *you* are the specificity of this type of ritual. This can be empowering, but it can also be terrifying, especially if you're new to this work.

Obviously, I have opinions about ritual, especially gender ritual: so many opinions that I even wrote this book. And yet my opinions are based on my brain, my body, my spirit, my context, my experience, and ultimately, what works for me. Your mileage may vary! This is why I have no illusions that my way is the best way, the most desirable way, or the way that will work for everyone. We are each working with our own set of experiences, preferences, abilities, cultural backgrounds, ancestral spirits, relationships, and neurobiology. Of course we will each experience these practices differently. Still, there are some things that tend to hold true across it all.

Some of the rituals in this book are meant to be short regular practices, designed to be done repeatedly or consistently. They may not fit the general idea of rituals, but I've included them because I find them to be very helpful for clients, and I come back to them again and again. The rest of the rituals are more involved: You need more than a few minutes to do them, and you probably wouldn't do them overly often. We covered helpful information for consistent ritual in chapter 2, so for the rest of this chapter, you can assume that when I say "ritual" I am referring to the type with multiple steps, meant to require some significant amount of time, and to be done once, or at least infrequently.

First things first: There are some parts of ritual that matter quite a lot, but usually in a big-picture way rather than a granular way. For example, in upcoming chapters I am going to teach you about cleansing, grounding, and protection. These skills are collectively called energetic hygiene. I strongly believe that energetic hygiene is nonnegotiable for ritual work in general. In fact, even if you don't do any other ritual work, energetic hygiene is an important foundation for life. However, the forms of energetic hygiene that work best for you will be yours to uncover through trial and error, and to practice. So, big picture, these practices are nonnegotiable, but in a granular sense, there's no reason that my version of these practices should be the same as yours.

The prework, or step 0, is preparing for the ritual and preparing the space according to the specific needs of the work you're doing. After that, many rituals will involve the following steps:

- ritual cleansing of the space and the people in it*
- warding and protection*
- building an altar and making offerings
- raising energy

- meditation or working with energy
- symbolic action
- divination
- gratitude*
- closing out*

Just as step 0 is preparing the ritual space and items, the final step is putting everything away. While each of these steps is important to me, the ones marked with asterisks represent what I consider to be non-negotiable. Cleansing, boundary and protection work, and then gratitude and closing out are like the bookends on either side of the ritual. They hold it together and help you to show up in a way that is both appropriate and powerful.

Yet the rest has a bit more wiggle room. An altar is almost always helpful, if not required, but what constitutes an altar can vary. It could be a candle and a picture, or some intentions written on a paper with a leaf from outside. It could also be an elaborate affair, prepared in advance, covered in offerings of fruit and baked goods and flowers. I've seen folks create a digital altar, and for folks who lack physical space or resources, I always recommend an energetic altar. My dear friend and collaborator, Zel Amanzi, talks about their body as an altar, particularly when working with the transcestors.

Raising energy is also quite elastic. You may wish to move, to dance, to sing, to clap, to do breathwork—there really is no end to the options. You could take this part out entirely, though I do find it to be very useful. Symbolic action is, perhaps, the most customizable of all. This could be about carving your name into a candle and lighting it. It could be a dance or rinsing your hands and feet or body in water. It could include moving items from one place to another. Really, the world is your oyster, and you will see many different iterations of symbolic action in the second part of the book.

Divination is probably the most optional of the steps, in part because it's deeply personal. Some folks prefer tarot or oracle cards, others runes or bones, or even pendulums. Some folks are terrified of divination and don't want to work with it or even consider it. For that reason I almost didn't include it as a step, except that I find nearly all of my clients ask for this when we do ritual. So, if you have not worked with any form of divination before, feel free to skip this part. If you're not feeling it, or the ritual is too public, you can also skip it.

When customizing a ritual that is already written, you could start by writing out the skeleton of the ritual, and analyzing if any of the pieces don't work for you. For example:

Ritual for a Thing I Am Doing

- Clear the space, bring in protection, set wards (see chapter 4).
- Call in transcestors or helpers (see chapter 12).
- Build an altar and make offerings (see chapter 5).
- Write out intentions.
- Carve a candle.
- Intuitive movement to raise energy.
- Imbue candle with intention, light.
- Meditate with flame.
- Read cards.
- Give gratitude.
- Close out the ritual.
- Dismantle the altar.
- Remove wards.

Perhaps reading cards doesn't work for you, but you still want a chance to talk to the transcestors or do some form of divination. You could replace this part with using a pendulum, casting runes, reading with trinkets, bibliomancy, or any other practice. Or perhaps no form of divination feels good to you. In that case, you could remove that part of the ritual entirely. Perhaps you're a pro at the cards, but instead of intuitive movement, you would rather sing or chant. These you can easily swap out. The idea is that if action *x* doesn't work for you, and action *y* provides pretty much the same outcome, that's an easy switch. But if action *x* is something you're fundamentally opposed to or uncomfortable with, you can just take it out completely.

If you are a total beginner to all of this, I recommend doing rituals as written at least a few times. Each of these rituals has been carefully crafted, and each element is included for a reason. Even if there is something about it that seems odd to you, it can be worth trying to see how you end up feeling. For example, my first tarot teacher, Lindsay Mack, taught me to call in my guides for my readings. I didn't really know what guides were, and I wasn't sure I believed in them, but Lindsay said to just try it and see what happened. So I did, and I definitely noticed a difference in my readings. That was very early in my spiritual journey, but through that experience I learned that sometimes you just have to try things, and see what sticks.

Over time, as you perform rituals, you will begin to get a feel for what works for you and what doesn't. If you only ever do the rituals in this book, that's fine! But if you find that you enjoy this type of work, it can be deeply rewarding to iterate on what is written here, and figure out what works best for you. Do you thrive on meditation and guided journeys? Physical rituals with lots of movements? Do you benefit from dressing in a specific way for ritual? These are all things to explore with yourself as you gain more experience in ritual. Of course, there are also aspects of ritual that are important to avoid, so let's talk about cultural misappropriation.

Cultural Misappropriation

Cultural misappropriation, also often referred to as cultural appropriation, is rampant in spiritual spaces. In the United States, it is hard to find a spiritual, occult, or metaphysical supplier who doesn't sell white sage, for example. White sage is a particular variety of sage that is sacred to multiple indigenous groups across the areas where it grows, commonly called Southern California and Northern Mexico. Because Native American traditions are often viewed as inherently mystical by white Americans, and because white Americans so often assume we are entitled to anything and everything that sparks our interest or intrigue, white sage has become commodified in a way that has been devastating to the traditional gathering grounds of various indigenous groups.

Mainly white spiritual people have spread the idea that burning white sage has some unique cleansing power for everyone. But the reality is that burning white sage is just one part of a complex and deep relationship that indigenous folks have cultivated with this amazing plant. The rituals and the relationships to white sage are specific to the lineages that have cultivated them, and when the smoke is simply deemed "best cleansing" by white people who have no relationship to these lineages, it is a deep violation. Because of the entitlement and disrespect that white people have shown around white sage, poaching has devastated the wild populations and traditional gathering grounds.

As a result of this cultural misappropriation, I've seen white sage for sale in the most random places, like clothing stores or trendy gift shops. White sage isn't supposed to be sold in this way at all, and this context strays far from its sacred purpose. Heartbreakingly, it is safe to assume that any white sage you see for sale at a storefront has been poached, contributing to the devastation. This never needed to be the case—almost every culture worldwide has some form of smoke cleansing tradition, and you can make beautiful, effective smoke cleansing bundles yourself from plants that grow locally to you. Further, what good is a spiritual tool that is made in a way that harms and dishonors the Earth themself and the traditional stewards of the land?

This is just one example of how cultural misappropriation functions in spiritual spaces. The traditions and sacred practices of a group of people are taken out of context and adopted in ways that are often coopted by capitalism. Racist views about the inherent mysticism of various cultures are deployed to create a veil of intrigue around these practices, even as they are debased by people who neither understand them nor have the knowledge to practice them accurately or well. And to top it off, these same cultures are denigrated by the very people who profit from appropriating them, either overtly or covertly.

When you're tailoring your rituals, it is important not to add anything to them unless you're familiar with its provenance and purpose. Words, symbols, practices, instruments, songs, dances—each of these carries power. When you misappropriate them without context, you are doing harm to both yourself and the culture and people you're misappropriating from.

Let's come back to smoke cleansing. Cleansing energy and space with smoke is an exceedingly common practice. Instead of defaulting to white sage because someone else told you it was the thing to use, consider your purpose for using smoke to cleanse, and then find an option that is both effective and appropriate for you to use. You could burn many different types of incense. You could dry a bundle of rosemary, mugwort, juniper, lavender. The options are endless, and none of them require you to steal from indigenous cultural practices that you don't understand.

The truth is, you should avoid cultural misappropriation because it's harmful and it sucks. But even beyond that, anytime that we engage in cultural misappropriation, we are robbing ourselves of the opportunity for an authentic spiritual relationship or experience and are therefore creating the conditions for a shallow, anemic relationship to the world around us and the web of all things. I believe that, especially for white people, this is part of why we often feel so spiritually empty and therefore spiritually voracious in a way that is never satisfied. Our ancestors chose to sever their relationship with lineage to avail themselves of unearned race-based power and privilege. We must heal that wound, even if it takes many generations. That healing starts by cultivating our own relationships to our roots, to the land, to the earth, and not through making a pale imitation of the relationships of others.

You might, at some point, realize that something you thought was fine is actually culturally misappropriate. This has happened to me, and to many other folks I know. At that point, you don't need to be overcome with shame. You don't need to make justifications or double down and get defensive. Simply choose to acknowledge the mistake, ascertain what is needed in terms of restitution, and move forward with this new knowledge. To spiral in shame is to center yourself—that's exactly the kind of thing to avoid.

It can be alluring to think that you can find some definitive list or rule book about cultural misappropriation and therefore avoid it entirely. This mindset is a sneaky form of perfectionism, which is itself a white supremacist impulse, and I encourage you to notice when it pops up. When folks are in this mind-set, and again, I see this mostly with white people, they sometimes go and find someone from whatever culture they wish to appropriate from and "get permission" from them for whatever they are doing. This too is deeply misguided and only increases the violence of the misappropriation.

Only you can be accountable for your decisions, choices, and practices around culture, so you must learn to be real and honest with yourself. Not in a self-centered, self-deprecating way, but in a way that is genuine and focused on right relationship with the earth, the web of all things, and the people around you. Colonization and the subsequent destruction and devaluation of many cultural practices are to blame for this situation. Yet you have the opportunity to choose if you will actively participate in decolonization or if you will further the violence with either action or passivity. I hope that all who choose to engage with this text will also engage with the idea that they have a role to play in decolonization.

What If You Don't Have Access to Lineage?

So, if you read the previous section and are now feeling hopeless because you have no cultural connection that you're aware of, take a few breaths. Many folks' connection to their root cultures is disrupted for one reason or another. Perhaps you're a person with a multiethnic background and there's no one particular culture that your family practiced. Perhaps you were adopted by folks from a different culture than yours and were therefore cut off from your own practices. Perhaps your family was connected to a culture at one point but chose to assimilate in order to access safety.

I can't speak to every contour of every situation. Something like, for example, cross-cultural adoption is extremely complex and there are a lot of moving parts. Such an experience varies dramatically from the specificity of someone whose ancestors were colonizers. Instead of trying to troubleshoot every form of cultural disruption in one chapter of one book, let me offer a few tips that have worked for me, clients, and students in the past.

First, remember that you do, actually, have ancestors. If in doubt, you can reach to your bones, your blood, the minerals of your body, the parts and pieces of you that contain not a few recent ancestors but millions of ancestors. Ancestors who are human and mammal and reptiles and fish and trees and plants and even stars. We sell ourselves short when we see only the possibilities of the current moment, which is so fleeting. You are a part of this earth. There are so many iterations of life that brought you to this place, this time, this moment. Start from that place if you need a place to start.

If you live on a land that is different from the land of your more recent human ancestors, I recommend reading Robin Wall Kimmerer's book *Braiding Sweetgrass*. It was revelatory to me to realize that we can practice becoming indigenous to a place—not to take over or claim some authority to it, but to develop a relationship based in reciprocity, mutuality, and understanding with the land that sustains our lives and holds our bodies. And also, if you know that your people come from a specific land, you can research that place. The symbols, herbs, art, foods, spirituality, myths, folktales, and practices of that place can begin to become a part of your self-concept and spiritual practice.

Finally, find ways to engage the web of all things that are based in personal relationship and knowing rather than relying on the understandings and experiences of others to guide you. For example, perhaps you wish to invoke or evoke a certain animal energy in your ritual. Rather than typing "symbolism of the owl" into a search engine, I recommend searching the animal itself. If you query about the symbolism, you will get someone else's relationship to that animal, whether it is an actual cultural understanding or some random blogger's opinion that they present under the guise of "ancient understandings" or whatever. If you read, for example, a Wikipedia page about an animal, you will be reading a set of common observations from which you can start to develop your own conclusions.

I learned about this type of animistic research first through adrienne maree brown's book *Emergent Strategy*. This book, foundational to so many of my ways of being in the world, uses different natural phenomena to consider how we as human beings can move, live, gather, and thrive. This was my first real foray into animism and is part of how I was able to conceive of transness as a form of spiritual practice. The book *Undrowned* by Alexis Pauline Gumbs is another excellent primer on this type of learning, and is published in the *Emergent Strategy* tradition. Gumbs examines various marine mammals and uses their features and adaptations to expound on many facets of life, relationship, and Blackness. These books might also help you to find your way through the complexity of developing a relationship to the web of all life.

But make no mistake: The goal here is to develop your own relationships. Working through the lens of someone else's relationship to a thing, symbol, place, or practice can be interesting and instructive if it is done in a way that is mutual and appropriate. However, especially when you're still early in the process of building your relationships to the web, it is good to focus on yourself first. If you read a dry article explaining the habitats and social practices of the crow, what stands out to you? What resonates and in what ways? When you think about those details, and then you observe the crows around you, what do you notice? Are you seeing them in a new way? What happens if you embody crow energy? How do you move differently? Do any parts of you awaken?

Over time, you will create your own lexicon of meaning through the very personal portal of your experiences and relationships. This is uniquely yours, yet it will likely "rhyme" with many of the things others say about whatever being (or biome, or chemical process, etc.) you're working with. These moments of resonance or rhyming can be instructive as well. This process is gorgeous and life-changing, and there will also be grief as you realize how much has been lost to the violence of colonization and white supremacy. It is good to feel that grief and to recognize it for what it is. If you want to understand more about this, one excellent resource is the book *My Grandmother's Hands* by Resmaa Menakem.

While this section is a brief introduction to these topics, I hope that it has given you some of the tools that you need to approach these challenges. Unlearning the processes and tenets of colonization and white supremacy in spiritual space and topics can be a lifelong process. It's not about crafting some perfect, blameless version of a spiritual practice (remember, perfection is white supremacy). Rather, it is about continuously doing what you need to do to become more connected with integrity, accountability, and grace.

Choosing a Venue

In some faith traditions, rituals are always held in the place where people regularly gather, like a church, temple, synagogue, mosque, or grove. I have had clients who chose to do their renaming ceremony in their religious gathering place when that was a possibility for them. For some gender rituals, you may need a gathering space with lots of room for people. For other rituals, it may just be you, or you and one or a few beloveds. Making choices around this is highly personal.

The most important thing to remember is that when a ritual is yours, it is your own preferences that matter. If you're doing a ritual by or about someone else, then their preferences can run the show. But I have found that when folks allow someone else's preferences to dictate their gender ritual, the ritual loses power. What feels meaningful and important to one person may be neutral or actively distasteful to another person. If you allow someone else to bring elements to your ritual that are distasteful to you, the ritual will be impacted negatively by this.

For example, let's say that you feel it is important to have the ritual by a body of water, but your lover or friend or mom thinks that the ritual should be indoors. If you capitulate to keep the peace, the ritual will be that much less powerful, because the parts of you that knew you wanted it to be close to water will not be able to conveniently forget that. This splits your focus and intention and disperses it in a way that really isn't helpful, especially in a ritual context.

In general, life is a series of compromises. But in this instance, take time to really sit with yourself before you compromise. Ask yourself if you're really okay with the compromise, and honor the answer when it comes, even if it is inconvenient, because it really does matter that your whole self is on board with this process.

With that said, before choosing a venue, you will need to be very clear on what ritual you're doing and why. In the second part of the book, I will offer my own advice about whether or not a ritual is ideal for a social setting or with a crowd, and you can decide whether or not you agree and what a more social setting means to you specifically. "More social" might mean you and two good friends in your living room, or it might mean a small crowd of people in a grassy valley in the local park. So much of this is personal and elastic, but it requires being honest with yourself about what you want, what works for you, and what does not.

Once you've chosen your ritual and the number of attendees that feels right, I recommend thinking through the logistics. The name change ritual that I reference in the introduction started as a Zoom ritual due to the lockdown phase of the Covid-19 pandemic. In the virtual version of that ritual, each of us wrote the name or pronoun we were releasing onto a paper, and then washed or dissolved the paper in water. When I decided to do this ritual in person, and we chose to do it on the beach, I felt gross about putting paper in the ocean. Though it is biodegradable, it still felt like littering. So after some thought, we chose to get natural chalk and write on stones (there are plenty of stones on the beach, and they came from the ocean). This is the sort of thing to consider when thinking through a venue.

Read through the ritual as written, adjust as desired, and choose a space and place that suits the needs of that ritual. When doing this process, you may also realize that you want an officiant to conduct the ritual. While the word "officiant" sounds intense, your officiant can be whomever you'd like. A spiritual adviser, ritualist, or similar would be fine, but this could also be your friend, chosen family member, beloved, elder, or whomever feels like the right fit for you. I would come to this person with some advanced notice if you can, so they can understand the ritual and go over the details with you ahead of time.

There is no wrong choice with these details! It is all about what works for you specifically. So take the time to think through what you're hoping to gain from the ritual and what might need to shift for the ritual to be fully aligned to your intentions. And remember, if things don't go according to plan or don't feel right to you, you can redo the ritual in the future.

Conclusion

There is nothing wrong with doing the rituals in this volume as they are written. In fact, I've worked quite hard to make them accessible for exactly that purpose. However, if you're interested in tweaking and customizing the rituals, I trust that this chapter will give you some helpful tips or at least start your wheels turning in the right direction. Read through a ritual first and gauge whether or not you prefer to tailor it. Check in with yourself and ensure that you're not engaging in any cultural misappropriation or compromising your deeply held desires or values for the purpose of the ritual. Once you've made those choices, it's all about deciding who, if anyone, will be invited, and where the ritual should be held.

As we move into the next chapters about pre-ritual and ritual skill, keep in mind that you do not have to know all the answers or do everything perfectly for your ritual to be effective and powerful. Messiness and struggle are a part of the learning process. Over time, if you want to, you can try all sorts of different approaches and find what works for you specifically. If this is wildly different from what I outline in this book, that's actually not a problem at all. You learn a lot from disagreeing with someone and trying things your own way.

If you take nothing else from this chapter, take this: You will not benefit from cultural misappropriation. Regardless of your relationship or lack of relationship with your root culture, you can and will benefit from building your own embodied relationships with other beings, land, elements, processes, and energies. Exploring this can kick off a rich and rewarding spiritual life and practice, and help you to tailor your rituals to suit your unique needs.

4

Foundational Skills

ENERGETIC HYGIENE

Introduction

Regardless of what types of ritual you are hoping to do, and how you intend to use them, there are some foundational skills that will help you to be as effective as possible. You do not have to use each of these skills every time that you do ritual, but it is good to at least familiarize yourself with them. Note that each of these skills could be a book in itself. Most of these skills have multiple books written about them. These chapters are simply meant to give the information needed for folks to use the ritual instructions in this book, but if any of these pique your interest, I've created a list of relevant texts in the resources section.

I've divided these foundational skills into the next two chapters. This chapter, covering cleansing, grounding, and shielding, offers an overview of these skills. I have put them into their own chapter because they are so vital—not only for ritual but also just for moving through the world as a trans person. If all you get from this entire book is a small, short version of each of these practices that you do regularly, then the entire book will have been worth it. That's how much I believe in these foundational skills.

I recommend reading through this section, and ideally, practicing each skill, before moving on to the rituals that I've created for this book. As always, take what works for you and leave the rest.

Cleansing

Cleansing is sort of a pre-ritual ritual. In fact, cleansing itself can be, and in my opinion, should be, a ritual in itself. You will notice that some of the rituals in this book are almost entirely cleansing, such as the ritual for after being misgendered. But even for rituals that are focused on other activities, cleansing is a vital precursor to the work. There are cleansing rituals that involve many different steps, but cleansing can also be as simple as spraying yourself with an herbal hydrosol, doing a breathing exercise, or washing your hands.

Regardless of how intense or simple, cleansing in some way, in some form, is required for doing ritual. At its most basic, it is a way of letting yourself know that you are moving from one thing to another, from mundane life into the realm of the magical or intentional. However, cleansing can be about so much more than that. You may hear folks discuss "priming the vessel" and cleansing as an aspect of this. That's because when you come into ritual, it's ideal to have relieved yourself of the burdens of your mundane life as much as possible.

If you're new to the idea of energetic cleansing, think of it like any other form of personal hygiene. What if you went through the effort of washing all of your bedding and remaking your bed in the morning, but then spent the afternoon doing some sort of incredibly sweaty activity in a muddy field? You probably wouldn't get into your nice clean bed before stripping your muddy clothes and taking at least a cursory shower. Similarly, cleansing before ritual is washing away the energetic sweat, mud, and dust that we retain when moving through this world. In this way, we can come to the ritual space without distracting energetic detritus, and focus more fully on the task at hand.

Another advantage of cleansing is energetic sovereignty. This refers to a state of being fully in your own energy and not being overly impacted or swayed by the energy of others. We will discuss this more in the section on attuning to your own energy, but your energy is one of your greatest and most vital gifts in this world. No one else's energy is the same as yours. You have a spark that is all your own, and that spark can be utilized as a form of protection, a way of discerning your path, a method of finding your people, and so much more. Your energy is precious. It is a gift. Learning to work with it mindfully is practically a superpower in this capitalist culture.

When we are fully in our own energy, we are able to see clearly what is ours and what isn't. This is one of those things that is quite simple to say and unbelievably profound to experience. Other people can, often totally unknowingly, drain our energy and use it to process their own thoughts and feelings. Most of us have experienced this, for example when someone starts using a conversation with us to process their own unresolved feelings about their gender identity. When done properly and with consent, a conversation like this can be so lovely, so affirming, and a genuinely positive experience for all involved. But too often, these conversations require vast amounts of unacknowledged stress for the trans person involved, while the other person will feel like they are "just talking."

This is just one example of so many. I could (and may) write another book just about energetic hygiene, energetic boundaries, and other important energetic practices. But for the purposes of this book, understand that cleansing your energy is most supportive when it is something that you do consistently, ideally every day or most days in some form. But at the very least, cleansing before a ritual, even in a small, easy, simple, and fast way, is a must.

There are many different forms of cleansing, and there is interplay between the energetic and the mundane. For example, cleaning and resetting your house in a totally normal, mundane way can move a vast amount of energy and serve you well as a cleanse. In fact, this type of mundane cleanse is often a necessary precursor when you're doing big cleanses, like the bathing protocol that I outline later. But cleansing need not require a top-to-bottom deep clean of your living space. If pressed for time, you can simply add a little sea salt to your soap when you wash your hands.

It is also helpful to think about cleansing with the various elements. How does water cleanse? How does air cleanse? Fire? Earth? We often use water to cleanse our bodies, but remember that Earth uses all sorts of processes and webs of relationship to facilitate decomposition, which in turn feeds the soil and the creatures therein, which in turn feeds all terrestrial life. So in that way, Earth has their own cleansing properties that are available to you energetically or conceptually.

Similarly, fire offers its own cleanse. Sometimes the brush must burn to germinate the next generation of wildflowers. Sometimes we need to sit in a sauna and sweat our stress away. Humans across many cultures have purified the air with burning herbs or incense. Fire is used to purify and refine metals, to distill herbal essences, to release intentions written on spell papers. And these are just a few examples.

Air is also a powerful cleansing element. As a breathwork facilitator, I often use the breath as a form of air purification for the body. We breathe deeply into our lungs, opening up space that might have been closed off, and release energy, emotions, and memories that may have been trapped. Sound can air-cleanse the body—have you ever given a primal scream? Have you ever sung a song with a group of people? These too are air-cleanses.

Cleansing your energy may not feel natural at first, but it is a vital part of your hygiene, much like scrubbing your body or brushing your teeth. You can stack some energetic cleansing work with these forms of bodily hygiene so that your physical and energetic hygiene become connected, and you're more likely to do both. This might look like taking two deep cleansing breaths before you brush your teeth, or washing your body with cleansing herbal sugar scrubs in the shower. As with everything, start with small, doable practices and build from there.

So, without further ado, here are a few of my favorite cleansing options. As with all the advice in this book, you can adapt it to suit your own needs and situation. Remember, all of these rituals are meant to serve you, so feel free to adjust, mix and match, and play until it feels right.

A Simple Energetic Cleanse with the Breath or Body

Method

1. Start by finding a position and place where you are stable. I often do this standing, though you can do it in any posture that feels good to you. If it feels good, close your eyes. If it doesn't, just let them softly focus on the middle distance, or focus on one specific point, whatever helps you be focused and present.
2. Find your breath. You don't have to deepen it, just find it, let yourself become aware of your lungs filling and emptying. If being aware of your breath makes you feel bad, stressed, triggered, or negative in any way, instead try pulling your arms in toward you, as if gathering something close to your body, and then pushing it away from you, as if gently scattering something. If arm movement isn't available to you, you can use fingers or toes, or use your mind to imagine pulling one color of light (your energy) into your body, and then pushing another color (all other energies) away. Feel free to combine these in any way that feels good to you.
3. On the pulling-in (inhale, motion, imagination), you can either say or think, "I call my energy back to me."
4. On the pushing out (exhale, motion, imagination), you can say, "I release all energy that isn't mine."
5. Do this at least three times, or as many times as you want until you feel ready to be done.

6. If you would like, you can imagine the energy that is leaving your body going into the earth to be composted and recycled. You can also imagine it dispersing into the air or into the water, or vaporizing in the light of the sun. I sometimes add to this by saying, "I release all energy that isn't mine back to the earth, where it can be composted into sustenance for all living things." You can come up with your own phrases that work with the elements that you resonate with.

Releasing energy to the earth can be more effective when combined with grounding, covered in the next section.

A Simple Energetic Cleanse with Water

I think this cleanse is best done in the shower, though if a shower is not available, you can do a hand-washing version, or complete the ritual in a bath. It is nice to have a scrub or exfoliating cleanser for this. I prefer a salt scrub (most preferably a black salt scrub, or a sugar scrub mixed with charcoal). You can also use an exfoliating glove, washcloth, or even just your hands.

Ingredients

- soap
    - exfoliating wash of some sort preferred
    - salt scrub, black salt scrub, or sugar scrub with charcoal would be ideal.
- If your soap is not exfoliating, an exfoliating glove, loofah, body brush, or some scrubby thing would be great.

- Some place where you can bathe, a shower preferred.
    - In a pinch, you can use a sink and wash only your hands.
- water
- optional: olive oil, jojoba oil, or another oil you like for your skin

Method

1. Begin by giving yourself a general wash (a normal shower). If you need to wash your hair, this is the time to do it. Allow yourself to become aware of the energy that you wish to cleanse. You might imagine it as a gray cloud around your body, as dirt, or just as a specific color of light around you.
2. Put the soap or salt scrub in your hand. (Tip: you can put some liquid soap and salt in your hand if you don't have a premade scrub, or tip some oil in your hand along with salt for a salt scrub on the go). Starting as high up on your body as you can (it's not always great to scrub your face with salt, for example, and you probably don't want it in your hair), begin scrubbing your skin. Don't go too hard! Just scrub and say, out loud or in your head, "I release all energy that isn't mine." Keep saying this and scrubbing slowly and thoroughly across all parts of your body, top to bottom. Try to always be moving down and away rather than up and toward. The words can blend together and become more and more automatic as you go.
3. Once you've scrubbed your whole body top to bottom, rinse all the soap and salt away, starting with the top of your head. You can let the water run over the top of your head for a bit before you move to rinsing the rest of your body. Continue the words, "I release all energy that isn't mine." You can add, "May it be fully washed away." You can also add any other words that you like.

4. If you like, you can imagine the energy (the color or texture you imagined earlier) washing off you and swirling down the drain.
5. Finish your shower, get out, and dry off. If you like, you can use an anointing oil or herbal oil, then moisturize your entire body with lotion or similar while saying, "I call my energy back to me."

Note

If you have mobility or other issues that prevent you from being able to access this ritual as written, that's not a problem! It works just as well to have a caretaker bathe you while you say the words out loud or in your head. You can also just wash the parts of your body that are accessible to you. For example, I can't reach the middle of my back, and sometimes can't fully get the bottoms of my feet. Just do what you can. The intention is the key here, which is why this can also be a hand-washing ritual if you desire.

A Simple Energetic Cleanse with Smoke

Ingredients

- something that makes smoke: ethically sourced and culturally appropriate herb bundle, incense
- some safe way to work with the thing that makes smoke: an incense burner with a handle, a heat-proof dish to set your herb bundle in

Method

1. Set up your space as needed with regard to ventilation and a heat-proof dish or incense burner. You can do this either

clothed or naked, so get into whatever state of dress or undress you prefer. Note that if you are clothed, you may accidentally singe your clothes or get ash on them, so be aware.

2. Light your incense or herbs. Move the smoke from the incense or herbs over your body, starting with your head. You can use your hand to fan the smoke where you want it to go. Make sure to let the smoke hit your ears, stick your tongue out, and let the smoke waft over and under it. If any part of your body feels particularly energetically rough, let the smoke hit it longer, until it feels complete. If you want to say or think words, you can! Anything about cleansing or releasing energy should be fine, or you can use the words from the previous rituals, "I release all energy that isn't mine."
3. Pay extra attention to the parts of your body that get smelly—armpits, genitals—as these are places where energy gathers and gets stuck. Again, work your way from top to bottom, and do your best to let the smoke hit your back as well, and if you are able, the bottoms of your feet (it's okay if you need to sit down for this).
4. Once the smoke cleanse is complete, you can anoint yourself with oil or tincture, or use the energetic shielding exercise "A Basic Shielding Practice," detailed later in this chapter, to seal your energy.

Notes

- If you live in a place where the smoke alarms are sensitive or you share your space with pets, you may want to skip this altogether and instead use a cleansing hydrosol (a water-based herbal spray) or the breath exercise described earlier.
- If you still want to do this, I recommend doing it in a room with an exhaust fan, such as the bathroom, or opening as

many windows and doors as feasible. Getting a cross-breeze through your space is particularly helpful. You can also run an air purifier while you do this ritual. I do all of the above to avoid causing health issues for my pets.

- On herb bundles: It is common that folks purchase and use white sage for this. Using white sage is unethical unless you are from an Indigenous culture that has an ancestral history with white sage (see "Cultural Misappropriation" in chapter 2). White sage is being poached at a devastating rate to feed the insatiable Western appetite for "spiritual tools" that can be easily found and purchased, in part due to the white Western fetishization of indigenous cultures. You can easily make your own herb bundles from plants that are appropriate to your lineage or purchase them from ethical sellers. For example, juniper and cedar make lovely smoke bundles. Other sages (garden sage, desert sage, pineapple sage) can be used, as can lavender, mugwort, and many others.
- Similarly, many cultures worldwide have smoke-cleansing practices. You can call this "smoke cleansing" or "smoke bathing" but don't call it "smudging." This word is culturally specific and should not be used to refer to smoke cleansing unless that is a part of your culture and lineage.

A More Robust Cleanse

You can combine these cleanses if you would like to create a much more robust cleansing protocol. Kaitlin Coppock has done something similar to create a bathing protocol that, while intense (it's over thirty steps long), is my personal gold standard for energetic cleansing. Her bathing protocol can be found on the Sphere + Sundry website under "Sphere Sundry Bathing Protocol" (www.sphereandsundry.com/bath).

Kaitlin's bathing protocol is beautiful, and she offers a robust explanation for each step. If you like this ritual, I highly recommend reading her version. I have been deeply served by this protocol. However, it's not the only one you can try; cleansing baths are ubiquitous across many traditions and require, at baseline, little more than salt and water.

This type of cleanse is meant for the times when you really need a hard reset. I like to do these types of baths on the new moon, whenever I have been through a particularly hard time, or before a particularly important ritual or devotional cycle. After you do this type of bath a few times, you will begin to understand when it is called for.

This is a time-consuming process. I recommend fully committing to that. Set aside a whole evening for it, maybe even a whole day. Take your time with the setup, the bath itself, and the aftercare. If you've never gifted yourself a day for this depth of self-care, give it a try and see how it feels.

Deeply inspired by Kaitlin's bathing protocol, here is a version that I use. This version of the protocol is included with Kaitlin's blessing.

Ingredients

- exfoliating scrub or scrubby washing tool: exfoliation glove, loofah
- nail clippers
- nail brush

- usual hygiene items (whatever you typically use to get clean)
- razor or trimmer, if desired
- shower
- bathtub or large basin; can even be a storage container that is clean
- incense or herb bundle (see notes on the previous ritual)
- incense burner with a handle or a heat-proof dish
- something to light incense with
- salt: sea salt preferred; black salt is good for cleansing and banishing
- bathtub additions: herbs, oils, tinctures, bath salts, honey, milk, beer; read the method for more details
- cup or pitcher
- herbal or sacred oil
- robe
- lotion or moisturizer

Method

1. Begin by cleaning the area where you will be bathing. Deep clean the bathroom, scrub the toilet, clean the floor and the countertop, but especially deep clean the bathtub or shower. You can also do this after cleaning your home or room (another form of energetic hygiene) and washing your sheets. The cleaner your space, the more the energetic reset hits.
2. Next, clean yourself as you normally would. I do the simple water cleansing protocol during this step—basically anything I would need to feel cleansed and prepped on a normal day. Give yourself

a very thorough scrub-down, and trim finger and toenails if needed. At least scrub under your fingernails and toenails even if they don't need to be trimmed. Exfoliate any rough patches, trim or shave hair that you would normally trim and shave, or if you want to ritualistically remove hair, this is the time to do so.

3. Once you're thoroughly clean, you can begin the magical or ritual part of this process, either right after or later in the day.
4. Begin by calling in your guides, transcestors, any animal or plant spirits that you want help from. Ask them to join you and to help you cleanse your energy and body.
5. Smoke-bathe your body using the instructions from the simple smoke cleansing ritual. Be mindful to get the areas where scent and energy collects.
6. Run the bath or fill your basin with hot water (not too hot; don't burn yourself). Add sea salt to your bath and speak aloud or think your intention to purify and cleanse.
7. Add any other additions, including herbs, oils, tinctures, and flower essences. I am listing many options, but it is best, at least at the start, to go simpler. In other words, sea salt is a must, but add only one additional item to start. One herb, one flower essence, one oil—whatever you want to focus on. Start simple, assess your preferences, and build your own practice.
8. Add beer to the water at this stage if desired. According to Kaitlin Coppock's bathing protocol, beer nourishes and insulates your energetic field. I quite like using beer for this, but if you prefer not to, no worries. If you prefer not to use beer, try adding some honey instead. You can speak over the water, asking your supporters (guides and transcestors) to bless the waters and to bless you through them.

9. Step into the tub, or step into the shower with your basin. If you are able, fully immerse yourself in the water at least three times. If not, just immerse the parts of your body that are able to fit in your vessel.
10. Take the cup or pitcher and fill it, then pour it over your head three times. Note, if you are doing this over a basin, try to pour the water so most or all of it falls back into the basin, and consider using a smaller cup.
11. Pour the water over each part of your body in this manner, at least three times. For example, three times over the neck, each shoulder, upper arm, lower arm, then hands. After you've done the upper body and torso, stand up and do the lower body, with extra attention to the genital region and feet. Be as thorough as possible, hitting every inch of your skin at least three times if you can.
12. Drain the water and rinse yourself if desired.
13. Get out of the bath, dry off, and anoint yourself with oil. This is a great time to pray a blessing over yourself. I prefer to write my own; some folks use blessings from their spiritual traditions. If you would like a blessing, here's one that anyone can use. Feel free to adapt it as you see fit.

Blessing

Now that I am clean, may the light of the [sun or moon, whichever you prefer—if using the moon, you can also name its phase if desired] allow all that I have released to flow back into the earth, to be received, composted, and to begin the cycle anew as fuel for all life.

I call upon the blessings of the elements. The blessings of earth, which holds and sustains my life and all life, which composts all that no longer serves and converts it into sustenance for all living beings. The blessings of the water, which flows through my body as it flows through the land, which washes away all that no longer serves and leaves me healed and purified. The blessings of the air, which fills my lungs and flows as oxygen in my blood, the wind that clarifies my thoughts and ideas and carries me through each moment. And the blessings of fire, the warmth of my body, the beating of my heart, the firing of my synapses, the light of the candle, the transmutation of the cooking fire. I call upon these blessings and more.

I call upon the blessings of my ancestors. May I be held, protected, and guided by my well and benevolent ancestors. May I act in ways that bring them joy, peace, and healing. [Feel free to add more from your personal ancestral practice, or if desired, remove this section altogether].

I call upon the blessings of my benevolent transcestors, known and unknown. May I bring joy, peace, and healing to all of my transcestors, and may I walk in the blessings and love of all benevolent transcestors. May my life be an altar to their dreams.

I call upon the blessings of all the benevolent spirits that watch over me and protect me, with deepest gratitude. I ask that you continue to watch over me and protect me, to assist me in my endeavors, and to bring peace and gentle satisfaction to my rest.

I call upon the blessings of those who joined me in the ritual—[name everything else you called in, such as plant spirits]. With deepest gratitude for all of your gifts.

I release all who joined me here, my heart and field full of your blessings and support. May I rest gently, my energy clear and prepared for my next steps. With deep gratitude, I release you, I release you, I release you.

A Note on Blessings

I have written the above blessing in an effort to give something broadly useful, however, pieces of it may not apply to you. I want to reiterate that you can adjust this as needed. If you have a relationship with, for example, one herb, then simply call on those blessings if that is more comfortable to you. The goal of this is "sealing" your energy with blessings after having cleansed it with this very intense form of bathing. And if the word "blessings" itself is triggering or challenging for you, I get it! You can use any other word that feels good to you. "I call upon the positive regard" or "I call upon the gifts" or "I call upon the love."

Final Thoughts

These are just a few of the forms of cleansing that you can work with, but my goal is that they will be adequate for you to begin learning about cleansing and seeing for yourself how helpful it can be. Remember, cleansing is most effective when you do it regularly, preferably daily. Quick cleansing can be most effective in the morning, after leaving or ending work for the day, or before bed. Pick the time that seems easiest and most doable for you and start there.

As for which form of cleansing to choose, I would go with whatever feels the most doable to you, especially for daily cleansing. If the breath cleansing feels easy and accessible, go for it! If a quick smoke bath feels good, go for that. As always, start small, simple, and easy, and go from there.

The larger cleansing rituals, especially the bathing protocol, are better for more occasional use. I use something similar to this protocol on new moons, before I do important things, or before larger rituals. You may consider it before any of the rituals in this book, before and after doing challenging things like seeing family, interactions with the legal or medical fields, or after a bad date.

Finally, it is very likely that if you begin doing this sort of energetic protocol, especially combined with things like grounding and shielding, folks around you will notice and will react. Similar to setting a boundary, this can be both highly instructive but also distressing, as some people may no longer wish to be in your life because they can no longer use you in the same ways as they did before. While I don't have space in this book to discuss this, there are several others who have. You can find more information in the Resources section at the end of this book.

Grounding

Of all of these skills, I feel that most folks will probably have at least a passing familiarity with grounding. Grounding is something that is seen across many traditions—I've personally learned about it in therapy offices, from meditation and mindfulness apps, in somatic workshops and trainings, and in the witch world. Roughly speaking, grounding is about connecting to the earth—the literal ground, or at least whatever spot you happen to be sitting in. At its most agnostic, grounding might be described as allowing your body to find safety in the present moment rather than being in the past or the future.

A lot of therapists will use grounding in this way—almost as an antidote to a trigger, or as a way to create more capacity for trauma therapies such as eye movement desensitization and reprocessing (EMDR). You may be familiar with grounding techniques such as the 5-4-3-2-1 method, in which you name five things you can see, four things you can feel, three things you can hear, two things you can smell, and one thing you can taste. This exercise is a way of bringing yourself into the present moment and grounding yourself in the reality of your senses and physical circumstances.

There are many more of these types of grounding exercises, and you may want to use one of them instead of what I offer. But for the purposes of creating and conducting ritual, our grounding practices are a little different. I see grounding as a form of rooting my energy to the energies that sustain me—for me, mainly, that refers to earth energy. You may want to root your energy into something else, but I would start with earth energy. Earth energy is particularly special to me, as a Taurus sun with several important planets in Capricorn (both earth signs). Regardless of your birth chart, earth energy is the energy of the body and the material world, and our connection to that energy is both vital and revitalizing.

I believe that our connection to the earth *is* our innate spirituality. Everyone has a connection to the earth, because we walk on it every day, we eat food that comes from the earth, and the food we eat was also sustained by the earth. When we die, regardless of how we are buried, we go back to the earth. Our bodies are made of earth stuff (and star stuff, but the earth is also made of star stuff), and our entire experience of life is filtered through earth. Human beings—all human beings—have an inherent relationship to the earth, and this is something that we can always turn to and return to.

Grounding for ritual purposes is about finding that point of connection and rooting into it, then using that root as a sort of stability or power to hold the ritual itself. This is particularly important if you are doing ritual in private or just with yourself, or if you are holding a ritual space for others. Whoever is holding the ritual space must be rooted in some way, and grounding is a great way to do that work.

One common way of grounding is to send a grounding cord into the earth. This is a great practice because it is customizable and personal when you get the hang of it, but it is also quite accessible to many different types of folks, regardless of beliefs or background. It is nice if you can do a cleansing activity before grounding. I suggest using the simple breathing cleanse from the previous section, then add this grounding piece at the end.

Sending a Grounding Cord

You can be in any position for this ritual practice, though if you're new to grounding, consider sitting in a place where you are well-supported and comfortable. Start by closing your eyes if that feels safe and comfortable. If not, let your gaze relax and soften in the middle distance. If you are adding this practice after the breathing cleanse, skip to the next paragraph. Otherwise, spend a little bit of time locating your breath. Become aware of the fact that your breath is moving in and out of your body. With your exhales, send out any energy that isn't yours, or anything that doesn't serve you. With your inhales, call your own energy back to you. If breathing this way doesn't feel easeful or good to you, you can use movement or visualization to imagine the inflow and outflow of energy.

Find the places where your body is connecting to the ground. This might be the bottoms of your feet if you're standing, or it might be several places along your legs, hips, thighs, and back. Find these points of connection and stability and focus your weight into them. When you feel fully supported, let your breath start to pool in the center of your body. If you're not working with the breath, you could instead move your hands to the center of your body, or visualize a ball of light glowing there. The light can be any color that resonates with your energy. Then, either with your breath, your movements, your visualization, or all of the above, begin pushing that energy, light, or breath down through those points of support on your body, as if you were a plant pushing roots down into the ground.

Start by imagining your roots going into the layers of dirt beneath the structure you're in. If you're on dirt or grass, push your roots directly into it. You can imagine your roots mingling with the roots of the plants and trees around you, or mingling with the mycelial network that connects all things. You can imagine your roots mingling with the little underground creatures if you want, pushing deeper and deeper until they hit ground water, then bedrock, then pushing easily through the bedrock. With each layer, your roots are able to spread out, even just a little, as well as going down. In this way, your roots create stability.

Push farther down through the earth's mantle, and then all the way into the core of the earth. If there's anything (energy, experiences) that needs to be absolutely vaporized, this is the place to release that thing. I like to imagine my roots really wrapping around the center of the earth and rooting into it. Then, leaving the root there, pull your awareness back up the root system, back through all the layers of the earth, and back into your body. Now you are grounded into the earth.

One pro-tip: At the start of this practice, you can simply release any previous grounding cords so that each grounding cord is fresh. I learned this from Sarah Faith Gottesdiener's class on protection magic. It is such a simple step, but made huge positive differences in my grounding practice.

Grounding with an Already-Grounded Being

Another variation is to ground with another being that embodies the type of grounding you're hoping to do. For the purposes of this exercise, I often connect with a tree, but you could apply this same method to a boulder, a mountain, any plant life, or just to lying on the ground itself.

Start with a cleanse—I recommend the simple breath cleanse, but do whatever cleanse feels best to you. When you're ready, connect physically or energetically with the tree or whomever you're connecting with. I like to press my palms to the trunk of the tree, or lean against it so that it supports my whole weight. If you are grounding with the ground itself, try to lie down in such a way that your weight is as evenly distributed as possible. If that's not available to you, just connect with whatever part of you can feel the ground and root into it. If you're working with something less robust that can't easily support those types of touch, you can connect with it in whatever way is mutually beneficial, whether that be by touch or energy.

At first, just let yourself feel it. Pay attention to your own energy and whether or not you're feeling it shift. Do you notice anything at all? Any shift in how you feel or what you're experiencing? Then, visualize or connect to the ways that being is rooted and grounded. For a tree, you might visualize or otherwise feel into what the root system looks like, how it spreads, and how it supports the tree. If it's the ground itself, you might feel into the thousands of small root systems that hold the soil, the water table beneath the ground, or my personal favorite, the bedrock. I often pray to the bedrock to teach me how to maintain stability even as the earth shakes around me.

See if you can allow your energy to match the rhythm of that being. Some might call this coregulating your energy with the tree or the ground or the boulder. You can do this in an energetic-only way, but if that doesn't speak to you, then try to develop some relationship to that being that works for you. That could look like contemplating what the scale of time looks like to that being. It could look like considering the various webs of relationships that being is in. It could look like considering your impact on that being and its impact on you.

For example, if I am connecting with a boulder, I could experience the feeling of it on my skin. I could feel into the energy of the rock itself. I could contemplate the way that time interacts with and on that boulder. I could imagine the time scale of my own life by comparison. I find connecting with stone to be particularly helpful for zooming out from my own anxieties and focusing on something large enough that my anxiety can't touch it.

When you feel complete, send gratitude to the being that you're working with. It is also good to leave a small offering. Some of my favorite options are water, my own hair, small flowers, rocks, or shells that I find along the way, or other natural items. Don't leave anything that would disrupt the biome of that creature or that would be construed as trash.

Final Thoughts

You can make grounding as elaborate as you want to. You can adjust or shift or add to the practice as I have listed it here in any way that works for you. Your grounding work can be a two-minute practice or a twenty-minute practice. Similar to cleansing, you get the most benefit from grounding when you engage in it daily, or at least most days. This is because over time, you learn what works for you. You also get the benefits of being more grounded in your daily life, and you develop a personal relationship to grounding that supports you, specifically, as an individual.

By doing grounding work daily, or at least consistently, when it comes time to do your ritual work, you will be well-versed in grounding. And it will prepare you well for our next foundational skill: shielding.

Shielding

Shielding is an especially important practice for trans and gender expansive folks to learn and to use daily. I don't know when you will read this, but I am writing it in 2024 and 2025. Currently trans folks are facing a backlash that is quite profound in its scope and violence. Shielding is not just for spiritual applications—it is also quite helpful in a practical sense for protection. If you know that you are going into a space where you will be misgendered or deadnamed, shielding can be part of how you preserve your mental and emotional health. My own shielding practice is a daily affair (along with cleansing and grounding) and allows me to hold immense amounts of heavy emotional work without taking it into my body. Without this skill, I can't imagine how I could do the work I do.

I've also used shielding to help manage my social anxiety, particularly when I am in crowded or unpredictable social settings. Effectively shielding often means that people don't notice me or speak to me, which can help me move through space without harm or without being triggered. All my life I've been the sort of person that people just come to. Random strangers have always felt very empowered to speak to me, often revealing traumatic or scary stories about their lives and their pasts. This is energetically draining and sometimes traumatizing. My life changed dramatically when I learned how to shield my energy from this. Now this almost never happens to me, unless I am tired or forget to shield myself.

Shielding for ritual operates on similar principles to any other type of shielding. The main difference is that when shielding for ritual, you are creating a container that can hold whatever the ritual requires so that you, or whomever you are officiating for, can be fully vulnerable within that container. Knowingly or not, we often keep all kinds of walls and barriers up to protect ourselves in this world. One difference between that and shielding is that maintaining walls is typically done unconsciously, while shielding is very much a conscious practice. Shielding also creates a shell or barrier around us, while the unconscious wall-building generally happens within us.

With your unconscious internal walls up, you may not be able to fully open to the ritual as it needs to happen. But with the shield externalized and holding the space, you can find enough safety to release those walls and allow yourself to fully surrender to the ritual itself. The shield will help you to experience the full impact of your ritual. As with cleansing and grounding, the maximum impact of shielding comes from sustainable daily practice, which can be customized endlessly.

A Basic Shielding Practice

Ideally, begin with the simple energetic cleanse using breath or body, then the simple grounding exercise in the previous sections. Once you have cleansed your energy and established your grounding cord, you can begin pulling energy up from the earth to mingle with your own. If you would like, you can visualize this energy as a color that resonates with you.

On your inhales, imagine pulling the energy up through your roots. I generally do these practices standing, so I pull the energy up through my feet first and foremost. Breathe the energy up and throughout your body, one breath at a time, until your entire body is suffused with it. Again, if the visualization or the breath are not effective for you, you can use movement: Try touching your body one piece at a time, or lying flat out on the floor, your bed, the couch, or grass.

Once you've suffused your body with earth energy, you can imagine that energy mingling with your own. I like to use pink and golden light to represent my energy. Using your breath or your movements, pull that energy to the center of your body. You can place your hand on your stomach for this if you like. With the exhales or with your movements, slowly inflate that ball of energy and push it out from your body on all sides. You can experiment here—I like to push my energy just an inch or so farther from the farthest reach of my fingertips and in a sphere of that size surrounding my body. You can push your energy out as far as you want, or hold it close to your body like a second skin.

When you've established this egg of energy, you can begin to suffuse it with whatever you want. You can create an energetic bubble that is soft and comfy on the inside for you, but hard and spiky on the outside. You can call in transcestral energy, plant energy, animal energy, ancestral support—whatever feels supportive, protective, and good to you. You can put anything you want inside this bubble, anything that helps you feel safe, protected, and held. You can make the exterior of your bubble anything that you want as well.

For me, it has been life-changing to incorporate thorn medicine for bubble building. Thorns are sharp and can be painful if you approach them without proper respect. Navigating thorns without being hurt requires a slowness, a gentleness, and a care that I want folks to emulate when they approach me. Sometimes this looks like taking a hawthorn or rose tincture while I'm building up my bubble. Both hawthorn and rose double as medicine for the heart, so I invite them to heal me even as they protect me. My thorn practice includes keeping thorns on my altars, especially my protection altars. I pay attention when thorny or spiney plants appear around me, which is often in Southern California, and I contemplate their unique thorns and how those thorns work to protect and preserve them.

Once you have established your bubble, you are ready to move on with your day, cleansed, grounded, shielded. If you would like to test the efficacy of these practices, I recommend you do them daily for a moon cycle, ideally new moon to new moon, or longer, and pay close attention to what changes for you during that time.

Establishing more energetic sovereignty and boundaries can be life-changing, and the people around you often respond in a variety of ways. Folks who have benefitted from absent or leaky boundaries may be surprised or even affronted by your new energetic sovereignty. This can be a challenge, and the early part of your boundary journey may require your relationships to be reconfigured to some degree. However, strong boundaries also create more capacity for you to grow into relationships that are mutually beneficial, where boundaries are understood and respected.

Prentis Hemphill is one of my favorite folks who talks about boundaries and how to work within them. I recommend their book, *What It Takes to Heal* if you would like to learn more about how to manage the social side of shoring up your boundaries.

Shielding with Physical Items

Bubbles are an energetic practice, and quite an effective one. I've also found it helpful to work with physical items for my shielding practice, especially when I know that I am headed into a more challenging situation or a space that requires more shielding than my everyday normal. I enjoy working with the idea and practice of adornment for this purpose. The idea of adornment is a broad one and can cover many different practices. For me, adornment includes specific items that I put on my body as shields, energetic absorbers, or things that shift the energy around me and affect what and who I pull in or keep out. In our class Accessing Transcestral Wisdom, Zel Amanzi talked about adornment as redirecting the gaze and shifting the perception of the actions available toward you. Adornment can be seen as shallow or merely aesthetic, and part of its power lies in its underestimation.

In my practice, adornment includes anything that you add to the body. Most often, this is jewelry and clothing, but I personally include anointing oils and perfumes as well. Some folks use makeup as adornment, and I think that gender-affirming clothing and prosthetics could be used as adornment as well, though they don't have to be used that way. Tattoos and piercings are powerful adornments and can also double as gender-affirming practices. In my experience, a lot of folks first experience gender affirmation from a tattoo or piercing. You can also use specific body jewelry in a protective fashion in an existing piercing. For example, I have a pair of earrings made from hawthorn thorns. I often wear these when conducting ritual, as they provide energetic protection and help fortify my bubble.

If you have adornment items and you want to task them with bolstering your protections, I recommend a ritual to give them that specific task. This could be described as consecrating them, although the spiritual connotation of that word may not resonate with you, so simply use whatever word suits you. This process need not be too arduous. Gather the item or items that you wish to charge with protection. Cleanse them with smoke, or if smoke is not available, you can put them on a bed of salt overnight. When you're ready to charge them, hold each item in your hands and let it warm with the heat from your body. Ask it to protect you, and be specific. I like to ask that the item will protect me, that any unhelpful gaze will slide off of me, that no one who wishes ill toward me will even be able to perceive me, and that no harm will come to me while I wear that item. Thank the item, and store it in a respectful manner when it isn't in use; for example, try not to leave it just lying around.

Adornment is a powerful practice, and like the other practices in this section, we can extend it out to cover the ritual space. We discuss wards more in the next chapter, and you can think of them in a similar way: physical objects that mark and protect a space, that shift or change the energy, that filter the energy of the participants, that redirect the gaze, that change the available actions.

Neither bubbling nor adornment have to be an arduous practice. As always, start with small and manageable versions of these skills, and build up from there.

Final Thoughts

The topic of shielding and boundaries is a big one, and we've only touched on a few basics here. There are some resources that I recommend to anyone who is interested in understanding more and extending their skill sets in these areas. First of all, if you're interested more in the protection magic and warding, I recommend *By Rust of Nail & Prick of Thorn* by Althaea Sebastiani. If you're interested in learning more about magic, generally speaking, with energetic hygiene as a part of that, Aidan Wachter's book *Six Ways* is my most robust recommendation. Sarah Faith Gottesdiener offers quite a few episodes on boundaries on her *Moonbeaming* podcast, and her teachings on energetic hygiene were foundational to my understanding of it.

Finally, don't neglect the part of boundary work that is about your own internal self. Trans folks tend to carry a lot of trauma in our bodies due to the ways that we're treated in the world. This trauma can make it challenging to establish and maintain solid internal boundaries, let alone solid energetic shields. My journey with boundaries has involved reading books, listening to podcasts, many hours of therapy, a lot of failure, and a lot of magic. Every step on the journey has been valuable, but it certainly hasn't been a short or simple process. Even if it seems like an impossible feat to develop and maintain solid boundaries, I promise that with time, effort, and consistency, you can and will.

Conclusion

Cleansing, grounding, and shielding are the three-pronged foundation to an effective energetic hygiene protocol. These three skills are vital for doing ritual work, but they are also most effective when developed as a daily practice. In each of these sections, I have given simple exercises to develop each of these skills, and in some sections, I have included more elaborate options as well. The quick versions are great for daily practice, while the more elaborate are perfect for pre-ritual work, lunar cycle work, and any time when you need to be more cleansed, grounded, and shielded. In general, the more energy you put into these practices, the more impactful and effective they become.

If you're feeling lost and overwhelmed about where to begin, just start small. Something that takes less than minute, but that you do every day (or nearly every day), will be more effective than a thirty-step protocol that overwhelms you. There's a time and a place for the more intense, longer processes; I've tried to give enough information for folks who might want it while not overwhelming those who don't. The truth is that this part of the book, these three skills, could each be books of their own. You can use the Resources section to find more information on these skills, and if you're interested, I highly recommend doing so. Energetic hygiene quite literally changed my life, and I would not be where I am without it.

5

Foundational Skills

BASIC RITUAL SKILLS

Introduction

Similar to the energetic hygiene skills covered in the previous chapter, there are a variety of skills that are vital to the creation and execution of ritual. As always, there will be different skills for different lineages, and nothing that I say here is meant to contradict or challenge those lineages. If you already have cultural or ancestral methods for conducting ritual, defer to those in your own practice. For folks who need a little more guidance, and especially for those who, like myself, have been severed from lineage by ancestors who chose to align with whiteness and hegemonic religions, these skills are available to you.

In part 2 of this book, you will see that I reference these skills freely and frequently. Even if you do not take the time to practice these skills before you begin reading the rituals there, I recommend reading through them first so that you understand what I am referring to later. As with the previous chapter, there are many different levels of intensity around these skills. Over time and with practice, I have developed versions of these skills that feel energetically doable and sustainable to me, and others that are very long and involved and require more planning and care. If in doubt, try the simplest, least time-consuming version, and see how you feel. You can always try something more involved next time!

And finally, please note that each of these skills really could be its own volume. As with the previous chapter, these are brief overviews to get you started. I've included resources at the end of the book to help further guide your learning, if you desire.

Opening and Closing a Space

Opening and closing a space for ritual is a deceptively simple practice. The quick version of this practice can be as simple as calling in guides, benevolent transcestors, and protectors and then saying aloud that you're opening the space for [specific ritual name here]. Then, at the ritual's conclusion, saying aloud that you're closing the ritual, and thanking each of those you called. It can be far more elaborate, which we will get into. But even in the shortest form, there's actually nothing simple about it. Opening and closing a space is a form of *ritual container creation*, and this is one of the most important aspects of ritual, full stop.

Without a container, ritual can feel hollow, performative, empty. With a well-constructed container, even the most quotidian actions can become sacred ritual. I'll offer a story as an example. Due to intimate partner violence that I experienced in a previous long-term relationship, I used to struggle mightily with bringing up my needs, frustrations, or struggles with my current partner. In my brain, I would have a thousand and one reasons why it wasn't worth it, why my issues weren't a big deal, were better kept to myself. Over time, this created a lot of resentment in my loving, healthy, committed partnership.

Through couples therapy, my partner and I decided to create weekly check-ins. We set aside specific time on the calendar and didn't schedule over it. During the check-in, each of us answered the question, "What am I not saying, and why?" The ritual container of the weekly check-in became a space with enough structure to support my bravery. Over time, even though I was very scared, I learned how to say the things I hadn't been saying, and to explain why I hadn't brought them up. This helped my partner better understand the shapes of my trauma, which operated to keep me quiet and resentful.

The container of the weekly check-in also helped us both to bring things to the table when we were feeling regulated and calm. The fact that we both showed up consistently, and that our shared intentions were the health and longevity of our relationship, helped us to be able to hear one another even when challenging or painful topics were brought to the table. When the topic at hand was too much even for the container of the weekly check-in, we would take it to couples therapy instead. Sometimes, you need a competent third party to hold the container!

The weekly check-in container became a sacred space for the ritual of intimate vulnerability—only possible because together, we created a container strong enough to hold this practice.

Ritual isn't required to make a container, but it is grounding and helpful, and can strengthen a container and make it more robust. Creating a ritual around opening and closing a specific space solidifies the container and strengthens it. Spiritually and energetically speaking, the opening and closing of a container is where you can set parameters around who is and is not invited, how the space will be held, and how the energy within will be managed.

Steps to Opening a Ritual Container

Start by engaging the energetic hygiene practices from the previous chapter. Some form of grounding, cleansing, and protection or shielding on yourself is step one. Be as robust as is needed for the ritual at hand. Knowing how much grounding, cleansing, or protection is needed is a skill that you can and will develop over time. In general, if you're doing a ritual with others, more energetic hygiene is needed to ensure you're fully ensconced in your own energy. If you're doing a ritual alone, but the ritual is intense, requires a lot of energy, or is related to a big emotion, more energetic hygiene is needed. For a daily ritual or a less energetically taxing ritual, you can do a quicker, less involved energetic hygiene process.

Next, you need to repeat the energetic hygiene process for the space you're in. For this, I recommend cleansing the space first. If there is mundane cleaning to be done, do that first. If there are windows that can be opened, open them. Then, cleanse spiritually and energetically. This can be done in myriad ways, including with smoke or incense, with hydrosols (water-based sprays made of distilled herbs), floor washes, and by reaching out to the transcestors and keepers of the space and asking for their assistance.

Once the space is cleansed, you can create energetic and physical wards to help hold the energy and keep the integrity of the container while you do your ritual. Finally, you can call in helpful spirits and energies before a ritual for the purpose of grounding and protection. We will go over the skills in depth, so if you're confused about how to accomplish any of the steps in this section, keep reading!

Over time, you will become more skilled at opening and holding a space for ritual. Let yourself be a beginner. Let yourself experiment. The rest of this chapter will focus on providing a starting point for your ritual skill toolkit.

Cleansing the Space

Similar to your personal energetic hygiene, cleansing the space may be more or less robust based on the task at hand. If you're in a physical space where you do ritual regularly, like in your own home, it is helpful to have consistent practices to cleanse and protect that space. Similar to cleaning your kitchen or bathroom, you need some basic hygiene on a regular basis, and "deep cleaning" less frequently. If you're preparing for a ritual that uses a lot of energy or involves multiple people, you will benefit from more thorough energetic hygiene than you practice for your daily efforts. As with all of these practices, you can modify them based on your individual ability and capacity.

For ritual, I recommend creating cleansing wards in the four corners of the room. My favorite method for this is to use some sort of open receptacle, like a cup, bowl, or jar, and add salt water to it. Tap water with table salt is fine; sea salt is better. If you have access to seawater, that is my personal favorite. You can also just put salt in the four corners of the room if water or receptacles aren't available. Note that if you're doing ritual outdoors, be careful with salt or salt water, as it can badly damage the soil. Don't put salt directly on soil or plants or in a place where it will wash into soil or plants. Don't pour salt water onto plants or into soil.

These types of cleansing wards create a space for energy to be sent to when it is released during the ritual, like a filtration system. Both salt and water are purifiers, but you can also add protection and boundary-holding herbs to these wards. This can be done with tinctures, essences, or dried or fresh herbs put directly into the water.

A preferred method for me to cleanse a space, especially my own home, is to work with a cleansing wash. These can be very simple or elaborate, and are a fun place to experiment and get creative. I cover the basic method below, but I encourage you to experiment and play. What plants grow around you? How do they impact the energy of a wash? In playing with different ingredients, you can begin building a relationship to various herbs and other materials.

SKILL
Making a Cleansing Wash

Making a cleansing wash is a simple and versatile skill that can provide powerful results. When I make my own washes, I often use them to mop the floor instead of soapy water. But I will set aside an extra bowl of the wash to cleanse the threshold, doors, windowsills, and any other point of ingress in the house.

Ingredients

- water
- pot to boil water in
- sea salt
- dried or fresh cleansing herbs
    - Some good options include rosemary, dandelion, chamomile, yarrow, hawthorn, vervain, lavender, mugwort, pine needles, or passionflower. Choose one or two, especially if you're still learning about herbs.
- vinegar (plain white vinegar is best)
- isopropyl alcohol

Method

1. Fill a pot with however much water you think you'll need to cleanse the surfaces in your space, and put it on the stove to boil. As it comes to a boil, you can add a bit of sea salt and dried or fresh herbs, or anything that feels good. Let it come to a boil then turn off the heat and allow the mixture to sit until it's just warm (not too hot). You can use ice cubes to cool it down faster.
2. Drain the solids out with a strainer and discard. To the liquid, add a half cup of vinegar per gallon of water. (Note that vinegar can damage marble, stone, and waxed or unfinished wood, so be mindful of where and how you're using it). I personally add a little isopropyl alcohol to mine too for faster drying.

3. Use the wash with a mop to cleanse the floor from both mundane dirt and spiritual gunk. You can put some aside in a bowl, dip a rag in it, and use that to cleanse the thresholds, window sills, and doors, and anything else you wish to cleanse.

Notes

- You can say incantations or prayers over your floor wash or surface wash while it is brewing. I like to ask my protectors and guides to bless it, to clear away any energy that isn't helpful or doesn't belong in my space, and to protect and boundary my space.
- When you add various components of your wash to the pot, you can invoke their properties, out loud or in your mind. For example, you can invoke the cleansing properties of water or salt, the protective properties of rosemary, the boundary-keeping of yarrow. If you don't know much about the correspondences of herbs, there are many different resources that are culturally specific. For a robust resource, I recommend Rowan + Sage's Herbaria, an online herbalism school.

Shielding and Protection

Just like with personal energetic hygiene, it is important to ground the energy of the ritual and the space. For this, I recommend doing your own grounding exercise and then speaking your intention for the ritual aloud. If you're doing ritual with others, everyone can breathe together, ground together, and then speak your intentions aloud together, or one at a time. This is also a great time to build a communal altar for the ritual. We will talk extensively about altars later in this chapter, so just bookmark this thought.

A practice that is both grounding and shielding is calling in spirits and energies before a ritual. The grounding version of this is to call in the benevolent spirits of the space and the land and to make offerings to them. Offerings can be water, energy, gratitude, love, food, acts of service, and more. It is important to note that, especially if you're new to this sort of thing, it is important to call only on the benevolent spirits and to make offerings that are generally pleasing, like those above. Over time, you can tend to your relationship with the spirits of the land and the places where you live, work, and spend time. No spirit is obligated to be in relationship with you, so don't approach this as if you're entitled to their time and attention. Instead, approach with humility and with offerings. Let them tell you how to move from there.

Finally, shield and protect the space. This may also involve calling specific spirits and making offerings, but this time you want to call in your protectors along with the energy of protective herbs and perhaps animal friends as well. You can place items in the four corners to protect the space, like herbs, rocks and crystals, or talismans. There are many different forms of protection items, but if you're not sure where to begin, think about what represents protection to you personally, or what forms of protection to see in nature. Sharp objects are great for this (nails, ethically sourced claws, thorns), as are things that naturally hold or create boundaries. Protective items gifted to you from beloveds or family are also excellent for this.

It is also important to tend to the mundane aspects of shielding and protection. To the best of your ability, ensure that no one can unwittingly walk into the space. If multiple people are present, all should engage in these practices to their ability and comfort level. Once the space has been adequately cleansed, grounded, and shielded, you can officially open the container. This can be as simple as making a declaration: "Now that we have [name the actions done to prepare the space and the spirits called on], I officially open this space for [name the ritual]." To this, you may also wish to add a blessing such as, "May [the ritual] be effective, and may all those involved be blessed through their participation."

Closing the Ritual Container

When the ritual is complete, the closing process will likely be a little faster than the opening, but it is no less important. First, declare that the ritual is coming to a close. Express gratitude to yourself and to all who joined, human and nonhuman, for their participation. After offering thanks to the spirits, herbs, water, land, and so on, speak aloud that you release them from the ritual space. Dismantle the altar, then dismantle the cleansing wards. Salt water can be flushed down the toilet. If a toilet isn't available, flush it down the drain.

Make sure that you and anyone else involved release your energetic ties to the ritual and to one another. At the end of the ritual, everyone should be in their own energetic bubble and no longer attached to one another. I usually accomplish this by imagining the connections dissolving in a gentle, loving way. I find folks don't often think about or realize that their energy is scattered through the many interactions they've had with others, so taking the time to do this as a group can be very supportive to folks who are early in their ritual journey.

It is good and helpful to have water or tea and snacks available after the ritual for you and anyone else involved. Eating and drinking is both joyful and grounding, and having that time to relax and talk can help everyone integrate the ritual and also come back into their body. Above all, try to allow things to remain spacious. You do not want to rush to close the ritual container, because it will feel unsettling and destabilizing for the work you've just accomplished. Be gentle, be easy, be spacious.

Working with Intuition

I often use my intuition extensively in the process of creating and conducting ritual. This can be baffling to folks who are new to intuitive work, and especially to those who are more conditioned in a spiritual structure that involves strict rules and relying on someone else to mediate between you and the sacred. Not all such spiritual and religious structures are negative. If you have found your home in a spirituality that uses a holy book or a highly articulated ritual structure, you don't need to doubt that just because I do it differently.

These structured practices were not helpful or effective for me personally. Finding my way into a more intuitive practice was a huge part of being able to be spiritual at all. If you too struggle with organized religion or high-protocol practice, this section is for you.

First of all, let me get one thing out of the way. For the reader who is currently thinking something like, "Sounds great, except I don't have an intuition," I guarantee that you do. For a lot of folks in modern Western culture, our relationship to our intuition has been disrupted, to put it mildly, in many different ways for most of our lives. Many of us are distracted by social media and other forms of constantly available content. Many of us are worked to exhaustion just trying to keep up under capitalism. Many of us are dissociated from our bodies due to trauma of one sort or another, gender dysphoria, or the aforementioned capitalist hellscape. Being in touch with your intuition often requires some level of somatic (body) awareness. When being in touch with your body feels impossible, being in touch with your intuition can feel impossible as well.

The fact is that you too can work with your intuition. You can find the unique way that it shows up in your body and mind. You can practice and build it like a muscle. You can learn, over time, to hear it in deeper and more nuanced ways. This takes time and practice, and it is very much worth the effort. I go deep on building your connection to your body in the first few rituals, found in chapter 6.

Intuition shows up differently for each person. Some folks feel their intuition primarily in body sensations. When I pull tarot cards, for example, my hands tingle when I find the card that wants to deliver its medicine. Sometimes, intuition can feel similar to your thoughts. It can be ideas, images, bits of lyrics, or anything else suddenly popping into your brain. It can be noticing the same type of bird or plant in a new way as you walk through the world. It can be a "gut feeling" or a smell wafting on the breeze.

Some folks get confused about how to tell the difference between intuition and anxiety. In fact, this is the number-one question that comes up when I discuss this topic in classes and with clients. To some extent, it will be helpful to build this skill for yourself. But a good place to start is noticing that your anxiety is often urgent, worst-case scenario, sirens-blaring stuff. Your intuition is often quieter, less urgent, more consistent, and much less likely to alarm you. If you feel fear, you probably can't hear your intuition very well in that moment.

Of course, the exception is if you're actually in a dangerous situation and acting on instinct. I find that many times, my anxiety tells me I'm in a dangerous situation, but I'm actually just having a conversation, or walking down the street, or even sitting in my bed with the cats, having a quiet evening. Anxiety is funny that way. Grounding has been very helpful here: asking myself things like what's happening to me right now? Am I safe right now? Do I have a house? Do I have food to eat? And if so, am I actually in immediate danger? If I am actually safe, I will ground, cleanse my energy, and try again.

Over time, you will learn to tell the difference for your own self, your own body. I have had an experience of my intuition, spirit guides, and transcestors warning me that danger was coming. It was a profound experience and taught me a lot about the difference between anxiety and intuition. The story will be for another project, but the upshot was that anxiety has me making contingency plans for my contingency plans, while intuition has me resourcing myself exquisitely, such that I can meet the challenges that arise to the best of my ability. If in doubt, root into that difference.

As you build your intuition, you can bring it to your ritual space as well. For example, instead of reading the table of contents of this book, you could flip through until you feel like stopping, and then see what ritual you land on. Instead of doing the ritual as written, you could assess which parts of it speak to you, specifically, and then use them to spin your own ritual. Often, intuition will lead you to do things that feel nonsensical or strange. Try them out and see what happens! My intuition has led me in all sorts of unexpected directions, and in hindsight, they make perfect sense.

Above all, if nothing else, remember that intuition is a thing we all have access to—including you! It can be developed like any skill set, and however it shows up for you is perfect. Intuition is also excellent for creating and tending altars.

Altar Creation and Tending

While you can do ritual without an altar, I personally believe that they always enhance the experience and efficacy of both personal and community practice. You might belong to a belief system or lineage that has specific rules on altar creation, and if so, feel free to skip over this part.

Creating an altar is, in my practice, an incredibly elastic process. Some of my altars are ornate, covered in items, permanent or semipermanent installations. Others exist in unassuming tins or jars around the house, or are assembled and dissembled in the same day. My body is one of my altars, especially for transcestral work. My beloved friend and colleague Zel Amanzi speaks beautifully to this—how gender-affirming surgery scars and impacts of HRT on the body are all features of a beautiful altar to transness and to the transcestors. There is no wrong way to create an altar, no wrong way to engage with the altar—what matters is intention and attention.

If you're creating a physical altar, it is often best to start with a somewhat minimal setup. Especially if the altar is more or less permanent, they tend to become more and more crowded over time. If an altar is meant to be erected and taken down during the course of a ritual, you can go as minimal or maximal as you want. An overflowing community-created altar can be a gorgeous thing, but a few precious objects on a windowsill can be just as powerful, just as potent.

Many folks begin by representing the four elements—air, earth, water, fire. Various cultures have different approaches here, and your culture may have additional elements and ways of working with them. As always, I invite you to take what works from this section and leave the rest. For example, you may start an altar with a feather (air), a stone (earth), a small cup of water (water), and a tea light (fire). If you need your altar to be portable (so nothing that will slosh, and nothing that will catch fire), you could represent water with a shell or a blue stone, and fire with ash, a red or orange stone, or matches.

From there, you can add anything you want—pictures, written petitions, items of personal or ancestral value, talismanic or ritual items, flowers or plants, food, ethically sourced animal parts or leavings (antlers, snake sheds, molts), scarves or cloths, tarot cards, oracle cards, candles, or anything else that speaks to you. An altar is also a natural space to make offerings. Any of the things previously mentioned could count as an offering if placed on the altar with that intention. Many offerings consist of food, water, alcohol, coffee, plants or flowers, money, incense, and anything else that is valuable, delightful, and good.

If you're creating an altar for a group ritual, each person can bring something to add to the altar. These items can be taken back with the participants when they go, or they can be left on the altar if it's permanent. If items are going to be left, then the person tending the altar should be someone who knows how to tend to such items, and how to handle them when they need to be taken from the altar in a way that doesn't create inappropriate energetic bonds and doesn't promote waste. This book is not meant to train officiants or priestexxes; that work is better done in apprenticeship. If you don't have a person who can act in this role competently, then let the altar breakdown be a group project, and let each person take back what they brought with them.

Once the altar is created, this can be the center point of the ritual. It need not be in the center of the room, though it can be. The altar can be a point of focus, it can be a place folks return to again and again, and it can be a witness to all that occurs in the ritual. Altars are important anchoring points, and the offerings on them are important for the energetic exchange in the ritual. If we're calling on land spirits or other spirits, transcestors, ancestors, and additional other-than-human beings, we need to come with gifts to exchange for their efforts and energies. Those gifts needn't be material—gratitude is a gift, prayers are a gift, attention is a gift—but reciprocity is important and should always be considered in the realm of ritual.

With altars, you have a great deal of creative license. I encourage you to use an altar to really represent what you, specifically, want it to. Your altar may be covered in cut flowers or potted plants; it may be opulent and colorful or restrained and monochrome. Create the altar that serves your intentions and your reason for the ritual.

Sigils and Petitions

For certain rituals, it is helpful to have some sort of way to indicate an intention or request that you're calling in for yourself. For example, in chapter 6, we work with the idea of calling in pleasure, and for that purpose, we work with a candle. You can certainly carve the word *pleasure* into a candle, but sometimes our work is more effective when it is less straightforward. A common, simple practice in many magic traditions is to work with a sigil. Sigils are symbols that carry a specific intention or identity that can be easier and simpler to work with in a ritual or magical context.

There are many ways to create sigils, and some of my favorite teachers for this practice are Aidan Wachter (check out his book *Six Ways*) and Ivy Bromius, who has a wonderful primer on sigil creation available for free on her website (www.circlethrice.com/project/free-guide-to-witchy-sigilcraft). The basic idea for a sigil is that you either want to create a symbol that refers to a specific person (like a stand-in for a name) or a symbol that represents an intention. Some people use an alternative alphabet to create sigils. Ivy recommends the Theban script because "it has provenance, it looks magical, and it's not appropriative or in copyright."

A petition is, essentially, a request for something that you desire. For example, a petition could be a request for more connection to your body, more opportunities for pleasure, more ability to be present with your pleasure. You can address the petition to whatever you want—a transcestor or transcestors (see chapter 12), a deity, your future self—choose one source to work with and see what happens. The main skill with writing a petition, the intention statement for a sigil, is to place it in the present tense and word it positively. For example:

- "I want to be more connected to my body" could be "I am delightfully and fully connected to my body."
- "I want more opportunities for pleasure" could be "I experience pleasure abundantly and often."
- "I want more ability to be present with pleasure" could be "I am fully rooted and grounded in my pleasure."

You want to be both brief and descriptive, and you don't want to use words with negative connotation. For example, you wouldn't want to say, "I never experience a bad day"—a sentiment that has both *never* and *bad* in it. Instead, try something like "My days are beautiful and fulfilling."

I combined these two skills because most petitions can be turned into sigils, if you'd like, and I personally find the practice of crafting a well-formed petition statement to be meditative and clarifying. Experiment with the wording until you've created the most concise version of the statement. Then you are ready to make it into a sigil, or use it in its full form on a petition. If turning it into a sigil, start by removing all the vowels as well as the repeating letters. So "I am delightfully and fully connected to my body" would reduce to "MDLGHTFCNB." If you're converting your statement into another alphabet, like the Theban script, do so. Or keep the letters in English.

From here, you can play. Use the curves and lines of the letters themselves to combine them into a form, then refine the form into a shape that feels right to you. Ideally, it shouldn't look like letters when you're done, but rather like a symbol. Don't worry if you can't fit every letter into it, and keep refining until it feels complete. Once you've created a sigil, you can charge it. This often requires staring at the sigil, allowing your eyes to go in and out of focus, until it appears to move or squiggle, or just until you feel complete. After this, you can place the sigil under a candle, burn it, or release it in some other way.

Similarly, you can prepare a petition by getting your statement down, then writing it on a piece of paper. In the class Eternal Flame: Candle Magic and the Sorcery of Flame by Damien and Amaya Rourke, I learned to write petitions on square pieces of paper, with the writing itself in the shape of a circle. Damien recommended writing the petition without picking up the pen. When I use this method, I write the petition over and over again until the circle has no real definable words that I can read. To charge the petition, I place a candle over it and let it burn.

You do not have to work with sigils and petitions to do effective ritual. However, these are two simple skills that I've seen folks do across all sorts of spiritual practices because they are simple and effective. If you're curious about them, I recommend experimenting and seeing what happens.

Tarot, Oracle Cards, and Divination

Hopefully by now it's clear that most aspects of ritual are modular and customizable, and this is no exception. Not everyone is comfortable with tarot, oracle cards, or various forms of divination. If you aren't interested in these practices, feel free to skip this section. With that said, using divination in a ritual can be incredibly powerful and can add a unique dimension to the practice. For example, if I do a ritual to connect to my body, and then I pull cards to ask my body what it needs me to know, I may get information in that space that is hard to access otherwise. If I am doing a transcestral ritual and I pull cards, I can ask the transcestors what they want me to know, or for advice, and get an answer right then and there.

Tarot is quite beloved to me—I've been reading tarot for most of a decade, and have read for hundreds of people over the years. I took classes with Lindsay Mack, Julia Inglis, Sarah Faith Gottesdiener, and others. Tarot is currently a popular modality, and these days there are many books and resources that are specific to queer and trans people as well. You can check the Resources section for my specific recommendations. I particularly recommend Lane Smith's tarot history in his book *78 Acts of Liberation*. Further, a lot of modern tarot interpretation is based on the work of Rachel Pollack, a recent transcestor.

That said, tarot can be confusing and overwhelming, and many of the illustrations on the traditional Rider-Waite-Smith decks currently in circulation, illustrated by Pamela Colman Smith, can feel reductive, overly religious, cis-centric, and just too white. Many talented artists and collectives have created more liberatory decks, but some folks prefer to read with oracle cards instead. While tarot is a specific, constrained system, oracle cards play by their own rules. There can be any number in a deck, and they can operate on any logic that makes sense to their creator. You can make your own oracle cards with some paper and a pen or pencil, even if you're not an artist. Personally, I prefer to read with both—I mainly read tarot, and then throw in an oracle card or two if my hands tingle for them.

There are endless forms of divination that don't require cards at all. Every culture has their own forms of divination, and divinatory tools can be made from pretty much anything. Some folks work with pendulums, others with dice or buttons, trinkets or twigs, the *I Ching* or the stars and planets, coffee grounds or tea leaves. Honestly, the options are limited only by your imagination and intuition—whatever speaks to you, so long as it is culturally appropriate for you to work with.

If you're incorporating divination, I recommend couching it within the ritual. It should happen after the creation of the container, after the cleansing, grounding, and protecting, after the altar is constructed. It will be most effective if you are mentally and energetically fully held in the ritual container when you reach out for answers or information. I also recommend being specific with the wording of your question. This doesn't mean that you need to be specific in what you're asking for per se. For example, if I am doing ritual with or for a specific entity, I might ask, "What would you like me to know?" This is a wildly broad and open-ended question. But knowing that the answer I receive is coming from, for example, the transcestors specifically helps me know how to interpret the information I am given.

You can also ask for more specific guidance. For example, "What do I need to focus on during my top surgery recovery period?" Or "How can I best move through disclosing my new name?" You may also benefit from a specific spread, especially if you're using cards. A spread is a specific formation of questions or themes that you use to pull cards. Having cards in specific points can aid your interpretation. For example, a popular spread is a simple three-card past, present, future configuration. You can make your own spreads, or you can look them up online or in books. If you're going to use a spread in ritual, I recommend designing the spread ahead of time and bringing it with you to the ritual. You will have enough to think about without having to design a whole spread mid-ritual.

Only you can know if adding divination to your ritual will be helpful or appropriate for the outcomes you're seeking. If you are not already skilled with a form of divination, that's okay. Just make sure that you have the books or resources that you need to interpret the information you're given. If it feels too scary or intimidating, don't worry about it. Remember that ritual can and should work for you specifically. It can meet you wherever you're at.

Keeping a Record of the Ritual

It's very easy to forget the details of a ritual over time: what it included, what went into planning it, what your divination said, and anything else. It often feels so potent in the moment, then fades into the pleasant fuzz of memory. But especially for larger rituals, it's really helpful to keep track of what went into it and what happened afterward. For this reason, I recommend keeping a record of your preparation, what elements went into a ritual, what happened during a ritual, any sensations or experiences, and then returning back to it and updating the record with any pertinent happenings over time.

I have a dedicated journal for this sort of thing, but you might prefer having a digital repository or folding it into another journal practice. Some options to note down include:

- the intention for the ritual
- what you did to prepare
- any specific prayers or invocations that you used
- any petitions or requests that you made
- who was involved, if anyone besides you
- the rhythm or order of operations in the ritual
- offerings and altar items
- card spreads or other divination forms used
- any body sensations you noticed during the ritual
- any information or intuitive knowing that came during the ritual
- what happens in the weeks and months afterward

Of course, you can note anything that feels important or helpful, or anything that you think your future self should know or remember. It is especially important to note things that you can't interpret at the moment for your future self to reflect on. Let's say that in the days leading up to the ritual, you constantly see the color purple, hear the same song, or see crows. Each of these could be an important omen or synchronicity that your future self may benefit from considering, so put them in the record. To be clear, an omen is some sign or symbol that gives you information about what is happening or will happen. A synchronicity is when things line up in a way that is too coincidental to be believed, often in a way that confirms your work or something you're currently thinking about. Both can be highly instructive, and even if you're not sure what they mean or if they are "real," note them down for your future self's benefit.

Over time, your record of rituals becomes something of a record of your personal journey, and helps you to track where you've been and how far you've come. Having this sort of record can be a massive resource on multiple levels, not the least of which is showing you how hard you've worked over time to actualize yourself into the world. Gender exploration and transition can be a long and challenging road. There will be days when you feel very frustrated about it all, or when you need to remember your strength and resilience. Recording your efforts is proof, from yourself and to yourself, that you have fought for yourself. That you have loved yourself in a concrete way. That you have invested in yourself with your energy, time, and attention. That past versions of yourself have manifested your current iteration, that future versions of you are already looking back at you with pride and deep affection.

Conclusion

We've covered a few of the basic ritual skills in these chapters. Each one of these skills deserves a book of its own, and I've tried to provide more resources at the end of this book. If you're feeling overwhelmed by this chapter and the previous one, take a deep breath! You don't have to be a master of these skills to do any of the rituals in this book. The more ritual you do, the more these skills will come easily and naturally to you.

Whenever I start to get too rigid in my practices, I like to remember all of my grandmothers who, centuries ago, were making magic with the plants around them, with the bits of animal left over from meals or found while foraging in the woods. My grandmothers who had no access to an apothecary or a crystal store but who had intimate and deep relationships with trees and rocks and streams, with the ocean waves off the coast of Ireland, and with the medicine of the hawthorn tree.

There is no perfection in these practices. They are always only ever *practices*, helpful actions that we come back to again and again because they create scaffolds of support in our bodies and spirits, in our relationships with ourselves, with one another, with the spirits of all the things around us, and with our beautiful earth. Like the grandmothers, we reach for what is available to us—dandelions are profound medicine, a transcestral plant, and they grow in sidewalk cracks just about everywhere. No matter where you are, no matter what resources are available to you, no matter what you believe or don't believe, a ritual can support you.

So, as always, start small. Start with basic versions of these practices. Start with the quick version of opening and closing a space and build from there. Start learning your intuition with small daily practices—even just taking one breath before you go about your day or respond to a text or choose food for lunch can help you center and ground your energy more than you otherwise might have. A few items in a tin or on a windowsill can make an amazing altar. Pulling one card every day, or whenever you remember to, can teach you more about divination than any book. And hastily typing a few bullet points into your notes app can be your record-keeping as long as you need it to be. Start small, and build from there as your capacity allows.

That's a lot of skill building. Are you ready to start doing ritual? Let's go!

PART TWO

specific rituals

6

Foundational Rituals

Rituals for Connecting to the Body

No matter how you identify or what your gender journey includes, your body can be a tremendous ally. There is a great deal of social pressure for trans folks to view our bodies as "wrong." We hear this most blatantly in the idea that trans people were "born in the wrong body," or in the notion that gender dysphoria is the end-all, be-all of trans identity and experience.

Some folks do identify with the feeling of being born in the wrong body, and there's nothing wrong with that. If you are applying this idea to yourself, that's very different from the idea being applied to all trans people everywhere, or seen as a prerequisite for being "trans enough." Similarly, gender dysphoria is a reality for many trans folks (myself included). However, it is not a prerequisite for trans identity. You can be trans and not experience dysphoria.

I have run multiple workshops dispelling these notions, and introducing both trans folks and mental health practitioners to the idea that transition is enhanced through embodiment practice. Through this process, I've developed my own theory of this thing we call gender dysphoria. Put simply, though many of us experience gender dysphoria as the *mind* feeling disturbed by the *body*, in reality, it is *the mind interpreting the signals of the body*. In other words, it is not our consciousness feeling disturbed by the body. It is our body letting our conscious brain know that something needs to shift. Our bodies have a wealth of information that we can avail ourselves of, including, but certainly not limited to, what would feel good, or at least what would feel better than the current state of events.

Our bodies are potent allies in gender exploration and transition. If you are medically transitioning, your body must do the work of that. If you have surgery, it is your body that heals. If you're socially transitioning, your body often experiences vigilance or hypervigilance around name and pronoun usage. It also experiences the transformations of clothing and hair changes. Even if you're exploring and not making any steps to transition, your body is wise and well-informed about what feels good and what doesn't. No matter who you are, or how your journey goes, developing a better connection with your body is a great way to approach gender work at any level.

This section contains five rituals, but each are simple and can be used as daily or consistent practices. Each builds on the previous one, so you can start with the first ritual, building connection to your body, and when you feel about 85 percent competent with it, add the second, and so forth. Eventually, you will be able to work with all five on a regular basis. After enough practice, these rituals can become integrated into your daily life in a seamless way. Ideally, these rituals will become how you move through the world and through your day, though this may be more challenging for some than others. Remember that consistently increasing your connection to your body even 5 percent will make a big difference over time.

Building a Connection to Your Body

Your connection to your body is like a muscle that needs to be built slowly and gently over time. Unfortunately, many trans folks are quite disconnected from our bodies. In my experience, this disconnection often starts quite early in life and is mostly subconscious, such that many of us do not recognize the levels of our disconnection or dissociation until we try to actively get back into our bodies.

If this is your experience, coming back into your body may feel incredibly challenging or confronting. You might feel tremendous resistance or panic even at the thought of connecting to your body. I want to be very clear here: This is fine. It's easy to get really judgy or mean with ourselves when we feel resistance to something, especially if we feel like we "should" be able to do it. The fact is, we all live in a culture where connection to our bodies is actively discouraged. Dissociation through consumption, social media, nostalgia, and other activities is strongly encouraged, while we're simultaneously trained how to objectify our bodies for their productivity and ability under capitalism.

For trans folks in particular, dissociation from our bodies is often one of our primary survival skills. If your mind feels like dissociation is the only way to keep you safe, of course it will reach for that tool when you try to get in contact with your body! This is not a problem, and it's not something to be ashamed of. Rather, you can engage with your body very gently, moving at the speed of consent, and learning together how to form new patterns, shapes, and relationships.

To that end, this ritual or practice is about doing very small, very simple daily or consistent engagements with the body. The purpose is to start building a connection or a relationship, or at least building a foundation where such a connection can exist. This practice can be as small as you need it to be for it to feel doable. If you notice that you just never have time or energy for the practice, that means the practice needs to be even smaller and more doable.

For the purposes of this practice, don't worry about doing a bunch of energetic hygiene before you start. If you happen to have already developed an energetic hygiene practice, keep with it! But the main purpose of this mini-ritual is to break through resistance and get you started on a larger process of somatic connection. To that end, it will be better to keep it easy and simple.

Begin by wiggling your toes before and after brushing your teeth or some other daily consistent activity. This can be thirty seconds of toe wiggling. It's as simple as it sounds. For thirty seconds, wiggle your toes. Try to focus fully on your toes for those thirty seconds. Pay attention to how they feel, to the temperature around them. If you're wearing socks or slippers, see how your toes react to the sensations caused by them. If you're not, pay attention to the sensations of the floor, rug, or whatever else you can feel. After thirty seconds, you're done.

If you want, you can wiggle your toes for a minute or two before bed, or right after you wake up. If wiggling your toes is too intense, try instead wiggling and stretching your fingers. If that's still too much, try moving and relaxing all the muscles in your face. Experiment with different body parts and muscle groups and find one that you can tolerate connecting with, even if only for a few seconds. Start there.

When that feels more doable, you can start adding variations. You can go longer. You can connect with other body parts. You can really focus your awareness on that part of the body while wiggling. You can close your eyes or experiment with sensory deprivation while wiggling. When you're able to connect with your body on a somewhat consistent basis, like 85 percent of the times that you try, go ahead and move into the next ritual.

Building Allyship with the Body

Now that you've established some connection with your body, you can start guiding that connection toward allyship. In effect, this is like building a relationship with any being. If the toe wiggling and other gentle movements are akin to letting your body know that you're there, that you are opening up lines of communication, establishing allyship is setting the intention for what your relationship will be.

This ritual can work even if you don't have a lot of trust or connection built with your body yet, as it simply requires you to acknowledge that your body will be experiencing transition, whether medical or nonmedical. Ideally, this ritual would be bookended with daily or regular connection rituals from the previous section, but even if those feel challenging, acknowledging your body as your ally or teammate in transition can begin shifting a challenging or painful relationship to a more sustainable one.

Again, this ritual is meant to be brief and very doable, and does not require any specific materials. Begin by connecting with your body in a way that you've established and that feels safe. Toe wiggling is good here, as is putting one hand on the heart and another on the belly, or any other form of connection that you feel good about. Try to take a moment to breathe, or do anything else that helps you focus inwardly.

This next bit might feel strange, but that's okay. Give it a try, and if you dislike it, you can find an alternative. Either out loud or in your mind, speak to your body. You can be as informal as you like. You can say something like, "Hey, buddy, good morning!" Or any greeting that feels natural to you, like what you might say to a friend. Then take a moment to articulate to your body that you're going through a lot of changes right now, and that you hope to work together through that process.

If it feels available to you, take a moment to listen, and just see if anything comes up. "Anything" in this case means thoughts, sensations, textures, emotions, visualizations, sounds—really, anything. Some of this might feel like distractions, and it may indeed be. It's normal for the brain to go into protective mode and try to distract from what the body wants to convey. So just allow whatever comes up to come up, and as much as you can, don't judge it.

Finally, convey some gratitude to your body, your mind, and your whole self. You can end the practice with some stretching, toe wiggling, or anything else that feels good and offers a moment of positive physical connection.

Over time, after letting your body communicate with you, you might try telling your body your intentions for a day or time. For example, if you know you're going to try a new outfit, pronoun, or name, let your body know. If you know you have a hot date with a cutie, let your body know. If you know you're going to be stuck at work all day and it's gonna suck, let your body know. The point is developing a sense of camaraderie, like, "We're all in this together." Over time, this will strengthen your bond with your body, and you will be able to experience that sense of allyship or teamwork.

Building Trust with Your Body

It is very common that we feel betrayed by our bodies, regardless of gender identity. I find for many of my clients, there is a wound around puberty, when the body began changing without our consent, and we had to cobble together a set of coping strategies as a result. But we can feel betrayed for many reasons, such as a chronic illness diagnosis, trauma (of any kind), and self-judgment.

The good news is that building trust is doable, even when it feels as though we've experienced betrayal. In fact, I bet you already know how to do it, because I bet you've built trust with other beings in your life and maybe even rebuilt trust after it has been broken. Building trust is about showing up, being consistent, and making repairs when ruptures happen. Even if you're very skilled with these practices in your relationships, building trust with the body might feel strange or mysterious.

A great way to begin is to find your "yes" and your "no." I've led hundreds of folks in this exercise, and it's always interesting to see how many different ways the yes and the no show up for us. I ask this: Follow the instructions exactly the first time. If you want to experiment after that, be my guest. But this exercise is done this way for a reason.

Method

1. Begin by connecting with your body. I recommend using the grounding exercises from chapter 4, or doing some toe wiggling followed by deep breathing, or whatever else you've found works for you. Then start to ask yourself very simple, basic, obvious, *not emotional* questions to which the only answer is "yes." For example:
    - Do I live on [street]?
    - Is my hair [color] right now?
    - Is the sky blue?
2. And anything else you can think of. The key here is that the questions should be very obviously yes questions with no emotional charge. As you ask these questions, pay attention to see what is happening in your body. Are you hearing anything? Feeling anything externally or internally? Are you sensing anything at all?
3. When you've done this for a minute or two, switch to "no" questions. It is important that you not ask emotionally charged

or difficult questions. I recommend instead asking absurd questions, which will provoke a strong reaction without an emotional charge. Examples may include:

- Am I on the moon right now?
- Is the sky purple?
- Is there an elephant on the bed next to me?

4. And whatever else seems silly or absurd but obviously no. Again, pay close attention to your senses, and to what you experience with each "no." After a minute or two, whenever you're ready, go ahead and start asking one yes question, then one no question. Go back and forth like this as long as you'd like, paying close attention to the differences in sensation or experience with each yes and no.
5. By the end of this exercise, you should have an idea of what your yes and your no feel like. These are individual to each person. For me, I feel yes up in my chest, like a lightness or warmth, almost a singing or effervescent sensation. I feel no down in my guts, and it tends to be a dull feeling. Your yes and no may be similar to mine, or they may be totally different. There's no right way to feel these. What matters is that you can experience some consistent reaction in your body that clues you into a yes or no.
6. Once you have recognized these sensations, go ahead and start using them in your daily life. Pick a situation that is consistent but not emotionally challenging. This will be different for everyone, but one example might be choosing a beverage in the morning. Maybe you have tea, coffee, water, and juice. Take a moment to connect to your body, then bring each option up in your head and check your yes and no. Then follow the consent of your body.

7. Over time, you can start doing this work with more emotionally charged matters. If things are too emotionally charged, you may not be able to feel your yes and no. That's okay, just do your best. If you realize at some point that you trampled over your body's yes or no, apologize and try to listen more closely.
8. If you're wondering what this has to do with gender, think of it this way. Many folks come to me and are very confused about what they "actually want." They might be curious about hormones or surgery, but deeply confused on whether or not they really want these things. Learning to hear and respect your body's consent can help you find the best path for you with gender exploration and transition, but it can also help in many other ways. If you can hear your body's yes and no, you can tell when and where you need to set boundaries. You can resource yourself, or at least shore up areas where your resources are being depleted.
9. If you've spent a lifetime crossing your own boundaries and shutting down the messages from your body (like me, for the majority of my first thirty years), it can take quite some time and a lot of work to undo that damage. But it is infinitely worth it, no matter how you identify.

Building Love with Your Body

If you've managed to build connection, allyship, and trust with the body, you might just be ready to start building love with your body. This is a fraught topic, and many folks struggle with it. When I was in my late twenties and my early transition, the thought of loving my body was very silly to me. My first memory of actively hating my body was when I was around five or six years old. In the twenty-five years or so after that, I hated my body actively and passively, pretty much nonstop. There was always something wrong with it, some way it was failing me, some way it wasn't good enough. So when I heard people talking about "body positivity" or saying positive affirmations, it honestly sounded ridiculous to me.

For me, it was important to build first from that hatred and negativity to a neutral place. The goal wasn't "I love my body; it's so beautiful" or anything like that. It was simply "I have a body and that's fine." I might have stayed in that space a lot longer if I hadn't gotten sick.

When I was in my first year of my PhD, I became terribly ill. One of my chronic illnesses was completely out of control and my guts were wrecked. I'll spare you gory details, but it was a set of severe and horrible symptoms that were making it impossible to get through the day (or night, for that matter). I decided to try a very specific strategy for healing my gut, which involved weeks of eating nothing but tons and tons of soup. Not just any soup, but soup I cooked myself, from bone broth that I made myself.

I did this to try to heal my guts. I never expected that it would also begin healing my relationship with my self and my body. It turns out, spending hours of every day expending all my energy on caring for my physical body changed the way I felt about my physical body. Having to prepare and cart around gallons of soup everywhere dealt a huge blow to my perfectionism and anxiety. I was able to surrender to the fact that I looked ridiculous with my giant backpack full of soup, that I looked ridiculous eating soup in every class, that I was constantly running late because it takes ages to eat soup, and so on.

It took two years of eating in a very particular way to change my gut, and it was one of the hardest things I've ever done. Further, holding myself and talking to myself through the intense pain and symptoms of my illness taught me so much about the impact of how we speak to our bodies and how they respond. So when I talk to you about loving your body and loving yourself, it's not through the fluffy lens of bypassing your feelings or saying empty words. I forged these skills through the fires of immense pain, an exploding gut, and many hours of physical care labor.

So how to build love with your body? While there are many ways, the most powerful I've found is through something I call "love looping." Love looping is the act of performing acts of love for your future self, and then when you receive the benefits of those acts of love, sending gratitude back to your past self for that act of love. Note that you don't have to feel anything specific for this exercise to work. You are loving yourself through your actions.

An example: Let's say it's 8 p.m. and I want to sit in bed with the cats, but the garbage can is full. I know I can take the garbage out in the morning, but I also know my morning self will be rushed with all his other responsibilities. So instead, I take the garbage out with the specific intention that I am doing so as an act of love to my future self. When I wake up, and my morning is a bit easier because the trash is already out, I take a moment to send gratitude to my past self for that effort.

Over time, these love loops can work in reverse. You can know that your future self will be able to handle something, so you can rest now in that knowledge. But be careful with this. It's very easy for our present self to put too much on our future self. Instead, when you choose to let your future self pick up the ball, do so as a mindful and loving choice to both your present and future self, and then try to do something that will resource your future self as well.

Method

1. For this ritual, simply choose one or two small love loops. I find it helpful to phrase them as "I will do [*x* thing] now so that my future self can have [*y* thing]." Like, "I will do the dishes now so that my future self will have time to do yoga in the morning." Or, "I will cancel these plans now so that my future self can rest this weekend." Write them out if you need to or find that helpful. Complete the love loops by sending gratitude to your past self. Try this for a week, then reflect on your experience and how it felt. Do you notice more ease? More energy in your body? Any other changes?
2. Next, try mindfully working with love loops for a month or a moon cycle. Pay attention to how you feel. See what shifts. When you start working with love loops in small ways, they often grow larger and larger. Framing your actions this way is very powerful, and once you get the hang of it, you might even shift to asking yourself something more like, "Is this the most loving thing I can do for my future self right now?" before making any specific decision. You may also find yourself loving and appreciating, and even forgiving, your past self. Experiment, play, pay attention to how things shift.

Opening Your Body to Pleasure

No matter who you are or how you identify, your pleasure is sacred. When we hear "pleasure," it is often in the context of sexual pleasure. While sexual pleasure can be deeply profound and important, the pleasures of the world are endless to those who open themselves up to receive them.

I'm aware that things like pleasure are often seen as frivolous. Pleasure is not considered a particularly serious endeavor, and those who seek pleasure are often looked down on or even considered suspect. Yet, an idea that I like to sit with, especially when the world seems overwhelming and evil, is this: What is it that makes life worth living? And again and again I come back to pleasure. When I am feeling hopeless, it is not high-minded ideals or philosophies that bring me back to myself. It is the gentle pleasure of seeing a friend, or coming across a beautiful flower, or having a cat sitting on my lap, purring.

Loving yourself, as we explored with the previous ritual, is all the easier when your body is a vector for the many sensory or sensual pleasures of life. Staying alive is more possible when you are able to experience pleasure often. The pleasures that you experience do not have to be bombastic or take hours out of your day, although both of these types of pleasures should be a part of life regularly, I think. Rather, as with most things, small consistent engagement is more effective at shifting a life than big occasional ritual.

So how does one open their body to pleasure? First, I find it helpful to practice receiving. In our culture, we often think of receiving as the lesser position. People tend to enjoy giving gifts, giving help, giving all manner of things, and if they don't enjoy it, they at least prefer it. Giving feels powerful and has the distinct advantage of giving us a little thrill of moral superiority, or at least a feeling of moral "goodness." But receiving is another story entirely. Many of us have a lot of stories about what it means to receive—what it means about us, what it means about the giver. Many of us receive uneasily because we become instantly and deeply aware that we now "owe" something to the giver.

This mindset is distinctly a product of hyperindividualism in modern culture. After all, a plum tree doesn't feel greedy or bad for receiving the rain or sunlight that it needs to grow and produce fruit. It doesn't feel bad when the pollinators fertilize its flowers, nor when birds and squirrels and other creatures eat its fruits. The work of Robin Wall Kimmerer, especially her book *The Serviceberry*, will teach you much more of these tender lessons. For this ritual, however, you do not need to read any book. You need only to open yourself to what is already here.

Start with a glass of water or tea, juice, whatever you drink to keep yourself hydrated through the day. Look at the beverage and spend a moment or two sending love and connection into it. Then take a drink while holding the intention to receive. This is a deceptively simple exercise that I learned from Sarah Faith Gottesdiener. Many of us don't realize, until we try to do something like this, that we have any block whatsoever around receiving. When doing this ritual with folks, I've been told that they found their throat tried to close to the water, or even that their body went into a fight-or-flight reaction. If you're experiencing something like that, know that you aren't alone. Go slow, be gentle, practice opening yourself up just a tiny bit at a time. Can you receive even 2 percent more? 5 percent? Can you focus, relax your body, and receive even 1 percent more? Let that be enough.

Continue this type of practice in other areas of your life. I find it easiest to start with things that involve you receiving from yourself, or receiving from a nonhuman person, like a tree or an animal. Can you focus on receiving your food? The moisturizer that you use on your body? Support from the ground beneath your feet? Connection and happiness from a pet? This is the prework for the ritual for inviting in pleasure. Focus on receiving until it no longer feels dangerous. Then you can give this ritual a try.

Ingredients

This ritual is highly customizable, and you can do it with some ingredients or none at all. Read through it as it is and then decide what ingredients are right for you.

- a candle, pink or white if possible (I recommend a tea light or other small candle; you can do this with a larger candle if you have a safe way to burn it.)
- something to light the candle
- something to carve the candle (I use a ballpoint pen)
- anointing oil for the candle, if desired
- your favorite perfume, if desired
- body oil or moisturizer
- as many things as possible that you find beautiful and pleasurable: flowers, your favorite clothing, sex toys, lube, accessories and jewelry
- tea, water, or some preferred beverage in a glass
- spoon (preferably metal)
- journal and pen

Method

1. Gather the ingredients above and set them in an accessible place where they won't be disturbed.
2. Start the ritual by taking a cleansing shower. You can do the shower cleanse from chapter 4, and wash away energy that isn't yours or isn't helpful.
3. When you've finished the shower, go to the place where you've set the ingredients. Take a moment to "dress" the candle

(prepare it for use) with anointing oil, or anything else you prefer to dress candles with, such as herbs, then carve your name into the candle. You can also carve symbols or sigils for pleasure if you like.

4. Call in whatever supportive spirits you wish to, like guides, transcestors, and protectors. Say a few words asking them to assist you in receiving pleasure. You can ask them to open the way for more pleasure to come into your life, and ask for the capacity to receive it. While you call out to them, light the candle. Here's an example invocation. If calling out to guides, transcestors, and protectors doesn't feel good to you, you can replace that part of the invocation by calling to your future self, highest self, the universe, or whomever and whatever you want. As always, edit to suit your needs.

   Beloved guides,
   Benevolent transcestors,
   Fierce protectors,
   I call you here to me.

   Open the way to pleasure
   In this body
   In this life
   In my heart
   In my mind.

   Let me soften into receiving
   In this body
   In this life
   In my heart
   In my mind.

   Let pleasure flow through me
   In this body

In this life
In my heart
In my mind.

Let me recognize pleasure in every form
In this body
In this life
In my heart
In my mind.

Let me learn to receive pleasure
In this body
In this life
In my heart
In my mind.

5. After stating this, or stating your intentions, spend some time rubbing the oil or moisturizer into your body. As you rub it in, try to focus on the sensation along with your intentions to receive pleasure. If desired, add perfume, and spend time allowing yourself to really soak in the scent and appreciate it.
6. Put on your favorite or pleasurable clothing, or just stay naked if you want. Put on your accessories, or let them sit before you. Arrange the rest of the pleasurable items before you in whatever way you want, and spend some time focusing on each item individually. As you focus on each item, spend some time considering how it brings you pleasure, or what about it is beautiful. Then, spend time considering the ways you bring pleasure to yourself, and what feels good on or in your body.
7. Send that energy into your beverage. Let your eyes rest on the glass, and then let your gaze get soft. Try to send as much "pleasure" energy into the liquid as you can. When you feel

complete, tap the side of the glass with the spoon three times. Then, drink the beverage.

8. If you would like, you can journal about your experiences here, or you can close out and end the ritual. When you're done, remember to thank anyone or anything that you called in, and dismiss them with gratitude and blessings.
9. Over the next month, pay close attention to both pleasure and receiving. Try to find at least one source of pleasure each day, and spend at least a little time letting yourself feel it. Try to recognize places in your life where you receive, and track the feelings and sensations of that process. Become a student of how pleasure operates in your body, mind, and heart. Repeat the ritual if needed.

7

Rituals for Gender Expression

Introduction

In the previous chapter, many of the rituals were meant to be performed with regularity, and to shift your relationship to your body and yourself over time. Learning how to be in relationship with your body is vital for all the reasons we discussed in that chapter, and it is a great way to start getting in touch with the various ways that you desire your body to shift and change. In this chapter, we're focusing more on rituals that pertain to gender expression. While you can do these rituals as many times as you'd like, they are not necessarily daily or foundational.

We will cover rituals in this chapter for a variety of gender-affirming experiences; the first four rituals are for hair-related gender expression, the last two for clothing- and accessories-related gender expression. Although we're covering hair, clothing, and accessories, there is no end to the things that you may find gender affirming, and I encourage you to experiment with using the ideas from these rituals to create your own if needed.

We start with the rituals for hair. Hair is a remarkable thing! The ideal arrangement of one's hair will differ from person to person, from gender expression to gender expression. Some folks want the hair on their head to grow longer and stronger; some folks want the hair on their body to go away; some folks want a beard; others want to banish a beard. There is no right or wrong way to approach the hair on your scalp and body regardless of your gender identity or expression, and these rituals can and should be customized to your authentic self.

One thing about hair is that it is a powerful symbol. In spell work, hair is often used as an ingredient because it has been on our body and with us for as long as it has been growing. Therefore in each of the hair spells, you will find an energetic component that you can use to release or cleanse energy as you release the hair, or a way to raise and imbue energy into growing hair. If you do spell work of your own, remember that you can always add that into these rituals, and use your hair as an ingredient.

Clothing and accessories are also powerful. The way that we choose to adorn our bodies is highly personal and impactful. Our clothing can telegraph information to others about us, our identities, and how we would like to be approached. Clothing can also be a powerful shield—blending in can be strategic, as can standing out and making a statement. The accessories that trans people add to our bodies can carry a deeper somatic and spiritual meaning than might meet the eye. A packer or a breastform can be more than just a piece of silicone. These items become a literal part of us for however long we use them.

When we begin using a new gender-affirming item or wearing a new gender-affirming outfit, we can use that moment to welcome this new friend into our lives and imbue it with the intentions we have for it. When that item or those pieces of clothing have served their function, we can release them with love and gratitude. If clothing or items are gently used, we can also donate them to clothing exchanges or friends.

Thinking through an animistic lens, these items are all beings—beings of cotton or spandex, beings of silicone or lace—all of them have their own energies and their own place in the web of all things. Pausing to work with these beings in an intentional way, in a ritual way, will help you continue building your relationship to your own body and to where you operate in the web of all things. Let's do some gender affirmation ritual, and see what changes in your life, mind, and body!

A Ritual for Your First (or Fortieth) Gender-Affirming Haircut

This is a ritual that you can do for each gender-affirming haircut, but it may be specifically important if you're going to cut off a lot of hair, get a gender-affirming haircut for the first time, or meaningfully change your look.

Ingredients

- scissors sharp enough to cut your hair (just a tiny bit of hair is needed)
- paper or something to write on
- pen or something to write with
- items that represent you, for example:
    - photos of you
    - drawings of you or representations of you
    - beloved belongings, like a piece of jewelry you wear every day
- items that represent the form you would like to call in, for example:
    - collage or artistic representation, as abstract or as concrete as you want
    - plants, dried or fresh, or even grown in pots
    - photos
    - figures of animals or humans
    - anything else that makes sense to you

- anything else you might like to add to an altar:
    - cup of water
    - feathers, leaves, rocks, sticks
    - items of significance
- oracle or tarot cards, if desired
- a candle (a tea light or chime candle is best for this; something that will not burn too long) and something to light it with

Method

1. Start by preparing a working altar or working space. This can be any flat space where you can put items and burn a candle. I like to delineate my working space with a specific cloth or piece of paper.
2. Take the time you need to cleanse and ground your energy and your space in whatever ways work for you, and set up any wards or protections you desire (see chapters 4 and 5).
3. On your paper or cloth, begin arranging the altar items. Put the unlit candle in the middle, in whatever receptacle you need to burn it safely (a candle holder or a heat-proof surface). Behind the candle, place the item or items that represent you. To the right of the candle, place the items that represent what you are calling in.
4. Keep the scissors, paper and writing implement, and cards close at hand, but don't use them quite yet.
5. Light the candle.
6. Take some time to connect with your hair as it is. You may wish to use this time to close your eyes and deepen your breath, or you may wish to move intuitively, or you may connect to your

hair in some other way. There is no wrong way to do this, but however you do it, try to take the time to consider how long it has been since your last haircut. Think through the life experiences that you have been through in that time, whether it's a few weeks or several months or even longer. As you can, try to send gratitude to your hair for all that it has seen you through.

7. Take your paper and write your name (your chosen name if you have one, your legal name if you don't) or personal sigil in the middle. On the left side, write down anything that you would like to release. Perhaps you want to release a behavior like people-pleasing, or perhaps you want to release the energy of an ex, or perhaps you're ready to let go of self-hatred or making yourself smaller for the comfort of others. Whatever it is, write it here. When you're done, place the paper in front of the candle such that the candle is roughly over your name, and your items that represent what you're calling in are over the right side of the paper.
8. Take the scissors in your hand and say, "As I cut this hair, I release [the things written on the paper]." You can say this three times if it feels good. Then, take the scissors and cut a small bit of hair, and place the hair over the things you're releasing on the paper.
9. Next, write what you're calling in, or what your intentions are, on the right side of the paper. If you want to, you can place some of the altar items that support these intentions over the writing on the right side.
10. Now sit or stand back just a bit and let your eyes go into soft focus, roughly on the candle flame itself. As the candle flame burns, try to let the paper before you become blurry. Let your eyes move slowly to the left, where you acknowledge what

you're letting go. Use the scissors to cut the paper such that the left side separates from the rest, but your name and the right side are still attached. Out loud or in your mind, say, "I release you, I release you, I release you," as you cut.

11. Then let your gaze shift to the right side. Try to let your name and the words written on the right side melt together as your gaze becomes softer and less focused. You can say, out loud or in your head, "I call in [the things written on the paper]." Again, say it three times if that feels good. If you want to, you can put a little spit on your finger and rub it into the right side of the paper while you say, "I call you to me, I call you to me, I call you to me."
12. If you would like to pull a card or cards, this is the time to do it. Journal or make a voice note with your reflections.
13. When you're done, fold the left side of the paper with the hair inside it and dispose of it, ideally in a trash bin that is outside of your home. If you can, keep the rest of the altar until your haircut is complete. When you're deconstructing the altar, thank each part of it before putting it away. You can take the paper with your name and intentions on it, fold it, and put it in your pocket during the haircut. Then, place it in a journal or some other space. When you have your next haircut, take the paper out and celebrate anything that is even 5 percent more present in your life.

Rituals for Shaving and Trimming

Shaving can be a profound practice. Some folks are taught to shave to remove body hair in order to fit more fully into beauty norms. Some folks are taught to shave as a right of passage into adulthood, or at least on the way to adulthood. Some folks love shaving—they love feeling the smoothness of their faces or bodies; they feel empowered through the process of tending to themselves in this way. Some folks are only shaving while they wait for a beard to come in enough to grow it out. Some folks don't shave their body hair, choosing to trim it instead. Still others have a complicated relationship to shaving, where they only shave because a younger version of themself was told that they must do so in order to be desirable.

And, of course, there are probably many other relationships to shaving and all that it represents. The same can be said for plucking, trimming, waxing, and any other way that we temporarily shorten or remove the hair that grows on our bodies. There's no wrong way to orient toward your body hair, and there's no wrong way to shorten or remove it, as long as you're being safe. But because of all this nuance, you may wish to customize these rituals quite a lot.

If you're unsure about customization, remember that you can always revisit chapter 3 for more pointers. But generally speaking, take what works from this ritual, and leave what doesn't. And don't sweat the details—temporary hair removal is, by definition, temporary. If something doesn't feel good or doesn't work for you, just try something else next time. Finally, you do not need to do a long ritual for each time you shave. For some folks, this is a daily experience. Instead, focus on what is most impactful and most doable for you specifically.

Ingredients

- whatever tool or device that you shave, trim, or pluck with
- some sort of cleansing option: a smoke cleansing stick, a hydrosol, or literally soap and water or alcohol
- shaving cream or soap, if needed
- any appropriate oil or moisturizer for aftercare

Method

1. Take a moment to briefly orient to the current moment. You don't need to do a big cleansing or grounding meditation unless you want to. This could look like one or two deep belly breaths. It could also be just closing your eyes for a moment and placing your hand on your heart.
2. When you're ready, take whatever tools or implements you'll be using and cleanse them in whatever ways you want to. This could be smoke cleansing, using a hydrosol, or cleaning them in whatever way aligns with the manufacturer's recommendations. As you do this, if it feels good, say out loud or in your head, "I cleanse you of all previous energy. I ask you to partner with me today for [whatever outcome you want]. I thank you for your sharp blades [and whatever else is relevant, like 'faithful gears' or 'helpful guards']. I ask that you take only what I ask you to take and nothing else."
3. Do whatever you need to prepare your skin and your shaving cream or soap. You can speak blessings over your skin and even ask your hairs for cooperation. For example, if you're plucking hair, you could say, "Follicles, please release each hair that I pluck neatly and easily." If you're shaving, you could ask your hairs to stand up straight and be cut easily, and that your skin stay strong and whole. Whatever makes sense to you is fine.
4. As you shave, feel free to use this as a little energetic cleanse. You can say, out loud or in your head, "With these hairs, I release all unhelpful energy. I release any energy that isn't mine. I release anything that is no longer needed."
5. When you've finished and you're using whatever moisturizer or skin care is appropriate, you can say, "I call my energy back to me. I call in the blessings of my benevolent transcestors. I call in [whatever it is that you want to call in]." Thank your tools and implements and products, and go forth with your day.

Ritual for Laser or Electrolysis

Laser hair removal and electrolysis can be important and impactful aspects of the gender journey for folks of all identities. It's common to need electrolysis before having either vaginoplasty or phalloplasty, but folks also have laser or electrolysis as a form of gender affirmation in itself. Laser and electrolysis can both be uncomfortable and painful, depending on the area being treated, the individual's pain tolerance, and whether or not topical numbing creams are used.

Pain is interesting from a ritual perspective. It can be an element that naturally deepens a ritual state of mind or one that creates an opportunity to practice different skills and perspectives. For example, I had a tattoo on my spine that I wanted covered up. The cover-up required a lot of intense shading directly on my spine, so it was quite painful. I focused on breathing steadily and used the session to meditate on the coverup process. I used the pain as a release for all that the original tattoo meant and for the person I was when I got it. At the end, my tattoo artist told me that I sat like a stone.

When we know that pain will be a part of an experience, we have a few options. We can tighten and brace against the pain, inevitably making it worse. Or we can relax, surrender, and allow the pain and see what opens up to us and in us through that process. Because pain is a part of the permanent hair removal process, we can make a choice about how we want to approach it. And that can begin with, and be bolstered by, ritual.

This section holds two rituals—a more robust, first-time ritual, and the follow-up rituals you can use as needed. There are some different ways to approach these rituals, depending on your specific situation and process. For example, we include putting on the numbing cream, but if that's not a part of your experience, you can do this same process with moisturizer (if your provider allows you to moisturize before appointments) or during the shower with soap.

For the First Appointment

You will need some time for this ritual, so I recommend doing it either the day before or beginning the ritual a few hours before your appointment. While the ritual itself probably won't take more than an hour, it's important not to be rushed.

Ingredients

- candle and something to light it with
- paper and something to write with
- moisturizer, numbing cream, or body wash, depending on your practitioner's recommendations

Method

1. Start with some energetic hygiene. Some cleansing, grounding, and boundary work would be helpful before this ritual. When you're ready, go ahead and prepare the space that you're doing the ritual in. If you would like to light a candle, this is a good time to do it.
2. Write down your most ideal outcome from the hair removal. Look at what you've written and check it for any negative language, then rewrite it in the most positive possible terms. For example, "No hair on my chin" could be rewritten as "Perfectly smooth chin." Then rewrite again, this time as if you're a future version of yourself who is celebrating this outcome: "I adore my perfectly smooth chin." This will be an anchor phrase that you use for this ritual, for the hair removal itself, and for future rituals. You can make more than one of these if you are getting

more than one body part treated, or you can make something more encompassing. You can also make new anchor phrases as treatment progresses.

3. Start by gently tapping your fingers over the area that you're getting the laser or electrolysis. If you want to, you can start talking to that area now, just saying hello and generally connecting with it. Take your numbing cream (or moisturizer or soap, if in the shower) and begin to rub it in while saying your anchor phrase three times. Then explain to your skin and hair what you're preparing for in as much detail as you can. Ask your skin if it can accept the pain and relax. Ask your hair to respond to the laser or electrolysis probe. Tell your hair that you're releasing it with love and care. That you know it's done its best to grow but that you need your skin to be smooth now.
4. If you're getting hair removal for surgery, allow yourself to imagine the best outcome of that surgery, and if it feels available, explain that outcome to your skin and hair. Add as much detail as you can—how good could you feel post-surgery? How much fun could you have with your newly configured body?
5. When you're done, thank your body for being on this journey. Thank your skin for accepting the pain and for healing quickly from any irritation. Thank your hair for leaving and not trying to regrow. Thank your whole self for doing what you needed to do for your embodiment to be more aligned.
6. Write your anchor phrase on a piece of paper. You will use it in the mini-ritual. Blow out the candle, and have a good appointment!

Mini-Ritual

This ritual is mostly meant to recall the previous ritual. If you're using moisturizer or numbing cream, you can certainly do this during that process, but this is meant to be doable even in just a few minutes.

Ingredients

- paper with an anchor phrase on it
- optional: candle and something to light it with

Method

1. Start by placing your anchor phrase in front of your candle. Then light your candle, if you're using one, and take three deep cleansing breaths.
2. Tap gently on the part of your body that is receiving hair removal, and say hello in whatever way makes sense to you.
3. Slowly, with intention, speak your anchor phrase over that part of the body three times. If you want, or especially if you're putting on numbing cream, and therefore have a little more time, repeat the requests from the first ritual to your skin and hair.
4. When you're done, offer gratitude to your body, to your skin and hair, to yourself. Blow out your candle, and have a great appointment!

Ritual for Hair Growth

Growing hair is just as gender-promiscuous as removing it. It's quite common for trans femmes and women to grow their hair out, or at least experiment with growing their hair out, but one of the most common fears that trans masculine folks come to me with is hair loss. Growing beautiful healthy hair is a top priority for many of us, and while sometimes genetics, chronic illness, and hormones make hair growth a challenge, we can do ritual to call in our hair growth and encourage our bodies to do what they can.

Now, admittedly, this ritual is built on a foundation of spell work. You don't have to believe in magic or be a witch to do this ritual, but if you are, feel free to infuse this ritual with any witchy elements that feel right to you. If working with deities or spirits is in your practice, this is a great ritual to involve them in. Venus and Dionysus are both deities that relate to beautiful hair and that are available to be in relationship with for anyone who does not have culturally specific hair-related deities that they can call in. If you're interested in learning more about Venus and Dionysus as explicitly trans deities, I highly recommend Charlie Clare Burgess's book, *Queer Devotion: Spirituality Beyond the Binary in Myth, Story, and Practice*. The specifics of working with these deities in an ongoing way are beyond the capacity of this book, and you are not required to develop a relationship with any specific deity to do this ritual.

This ritual also benefits from the addition of various products. I recommend using the shorter version of the ritual as a daily practice for inviting in hair growth while using hair-care products. If you're using topical hair growth formulas, this is a perfect thing to do while you apply them. If you're not using any particular product, the physical act of massaging your scalp will still be beneficial to your hair growth.

Ingredients

- optional: hair growth oil or product of your choice
- optional: scalp massager (you can use your fingers)

- optional: candle
- something to carve into the candle (I use an old ballpoint pen; you can use a candle scribe or a toothpick or really anything with a point)

Method

1. After gathering the materials and preparing your space, begin your energetic hygiene. Cleansing, grounding, and boundary work in whatever capacity you prefer is fine for this purpose. Prepare for this ritual by getting very clear on the results you're hoping for, and putting them into a few short phrases. Similar to the hair removal ritual, you will want to put these phrases into positive-only language and present tense. For example, not "I don't lose my hair" but rather, "My hair is full, thick, and beautiful" or "I am pleased with my hairline."
2. If you're using a candle: take the candle in your hand and close your eyes. Envision the result you're looking for. By "envision," I mean that you can picture the result, but if picturing things in your head is not available to you, you could also hold a description of the results in your mind, or let yourself feel what it would feel like to have those results. Maybe you would sit a little straighter, walk with a little more confidence? Try sitting that way now, or walking with that confidence now. See if you can feel it in your body. When you've done this, focus that energy into the candle in your hand.
3. Take the candle and carve into it your name and a word or two that represents your desire. I recommend the phrase "beautiful hair." If it is in your practice, you could also dress the candle with appropriate oils and herbs, including the hair-growth oil. If you're invoking a deity, feel free to add their name or glyph to the candle as well, and say any invocations or prayers that

you'd like. When you've finished, light the candle and set it before you. If working with a deity, feel free to use this time to place offerings on an altar or burn incense—whatever your devotional activities require.

4. Take your oil or hair-growth product and apply as the instructions suggest. Begin massaging it gently into your scalp with a massager or your fingers as you call in the results you're hoping for with the phrases that you made earlier. As you massage, you can also send love and affection to your hair follicles, to the skin of your scalp, to the hair that you currently have. This can simply be about literally telling them, "I love you. I see you doing your best. I'm so proud of you."
5. Massage thoroughly and do your best to focus on the outcome you want. Try to do this for at least five minutes. As you focus on the outcome, try to also accept and receive that outcome into your body. You can even spend the last minute or so of this process saying, "I receive, I receive, I receive."
6. When you're done, you can allow your candle to burn to completion in a safe way (never leave a burning candle unattended), or you can extinguish it and light it again each day when you do the abbreviated ritual. Send gratitude to your scalp, your hair, and your body. Send gratitude to your deities if evoked, and then release them. Wash your hands and put away your materials.

Abbreviated Ritual

Daily, or as often as possible, repeat the last three steps of the ritual above. Massage your product into your scalp, speaking your intentions over your scalp as you massage. Give special care to any areas where hair is thin or thinning. Send love to your scalp, to your hair, to your body. Receive your own love and intention. Light your candle if you desire.

Use this time as a time to really dedicate to yourself each day. If it feels possible, dedicate this time to your own pleasure, care, and embodiment. If you want to add skin care or something similar, you can! Taking even a few minutes to gently tend to your body each day is its own form of a beauty ritual.

Ritual for Releasing Clothing and Accessories

Many of my clients have a complex experience around releasing clothing and accessories that no longer serve. This can come earlier in social transition, like when someone realizes that their closet is full of clothing that they no longer wear and that they can't imagine wearing again. It can also happen when someone has had a surgery that eliminates the need for a binder, a gaff, a packer, breastforms, or some other gender-affirming item. Finally, it is very common that as we progress in our understanding of ourselves and our gender process, we may find our gender expression or style shifting, even after years. In any of these instances or in similar ones, it is worthwhile to use ritual to honor what we are releasing, the place those items had in our lives, and what we're moving toward.

This is a ritual that you might want to bring a friend or two in for. I've had clients who struggled to clean out their closets on their own. They would become overwhelmed with the scale of the project or the emotional baggage of the items they had to sort through, which created a negative feedback loop about addressing the issue at all. However, bringing in a friend or two helped with alleviating some of the overwhelm of the project in general, even if the friend just sat with them as they worked. I have written this ritual so it can be done alone or with a friend or two. Read through the ritual and choose the approach that works best for you.

This ritual is also particularly modular—adapt it as you wish! Some folks will be blessing and releasing a few items, while others might have to bless and release multiple bags, boxes, or containers of clothing. If you're releasing two hundred items at once, you may prefer to bless and release them by the container instead of by the item. You might also do the gathering and cleaning of the items as a separate process from the ritual itself. Choose what works for your specific scenario.

Finally, consider donating any items that are in gently used or good condition to a gender-affirming clothing swap. This can be a beautiful way to make sure that you reduce waste and allow other trans people to benefit from your clothing. Something that no longer works for you could be a gender-affirming dream for another trans person. If you're planning to do this, I recommend adding a little bit of t4t love to the items before you send them on their way. See the end of the ritual for a note on this.

Ingredients

- whatever clothing or accessories you're releasing
- optional: paper and something to write with
- items for energetic hygiene (see the ritual description)

Method

1. If you're starting from the process of cleaning out a closet, a chest of drawers, or a storage container, start by cleansing yourself and grounding yourself in your intentions. Engaging with all of these items from a previous version of yourself could be disorienting or upsetting, so take the time that you need to

really ground into your intentions for this process. If you need to, write down your goals and intentions and tape them to the wall, or put them in a place where you can see them. If you want to, create a playlist that reflects the kind of energy you want to bring in for this process. Get clear on your intentions and the energy you're channeling for this process, then energetically boundary yourself. Remember that you are who you are right now, and that all the previous versions of you helped you reach this point. No matter what painful or positive memories these items hold, you can release them with love, and hold yourself in love while you do so.

2. Call in friends or helpers if that feels good to you, ideally folks who understand the depth and importance of this process. If you're sorting through or processing a lot of clothing, you could consider separating it into pieces that have strong emotional attachment or meaning, which could be dealt with one at a time or with more intentionality, and pieces that have less charge and can be dealt with in bulk.
3. Once you've determined which items you're releasing and letting go of, follow this ritual either with each item, or with all at once.
4. Either touch or look at the items. Spend some time reflecting on what those items have meant to you, and who you were when you wore them. If the items brought you a lot of joy, or represent a specific person that you were during a specific part of your life, take some time to acknowledge that and to offer them gratitude. Consider who you were when you wore those items, and send that person love and gratitude. If the items were a gift, and you have gratitude for the spirit of that gift, if not the reality of the items themselves, spend some time sending that gratitude to the giver.

5. When you feel complete, speak aloud, "To each of you, I thank you for all you have given me. I send you on to be a beloved and affirming companion to someone else in the future. While I retain my memories, I release you, I release you, I release you. I cut all ties from my energy to yours, and I release you into the world. With blessings and care, I release you."
6. At this point, feel free to use energetic hygiene methods to release the items. This could include:
    - ringing a bell over the items and over your body to break up and release energy
    - a dance party or intuitive movement to music
    - using smoke cleansing (be mindful not to get ashes on the clothing)
    - spraying yourself down with a hydrosol
    - washing your hands with salt and soap

    Or anything else that feels appropriate.
7. When you're done, if you're sending these items on to other trans people, feel free to add this blessing. Take time to imagine how someone would feel about these items if they were really delighted and affirmed by them. See if you can bring a lot of love into your heart for that person. If desired, you can rub your hands together briskly to generate warmth, then place them on your heart and feel the warmth as love for this imagined trans person and all that they represent.
8. Push the warmth and love from your heart into your hands, then put your hands on the items and imagine the love and support flowing from your hands into the items. If it feels good, say something like, "I pour these blessings (or this love) into you." Continue until all of the items have been blessed in this way.

Ritual for Blessing New Clothing and Accessories

Gender-affirming clothing, accessories, and prosthetics can mean many things to us. Receiving these items and integrating them into our lives or wardrobes can be such a joy. Seeing your chest or crotch flattened for the first time can be a joy, a thrill, a moment of great peace, but it can also surface a lot of grief, or result in sensory overwhelm. Wearing gender-affirming clothing can feel incredible, but it can also throw all the ways our bodies feel dysphoric into sharp relief. I might finally get the pair of jeans that aligns with my identity, and they might fit me just fine, but my hips look all wrong. Or the shirt I want to look so hot in emphasizes my chest or shoulders in all the wrong ways.

Receiving gender-affirming items into our lives, wardrobes, and rotations can be nuanced and complex, and using ritual as one way to process this experience can allow us to open to that complexity without automatically assuming we are weird or wrong for feeling all that we feel.

The goal of this ritual is to make space for whatever it is that you are experiencing, and also to bless the items that you're bringing into your life and to be explicit about what your collaboration with these items will include. We approach each item with an explicit acknowledgment that it carries its own energy, and we make clear what we want our relationship with that item to be.

You're invited to create an altar as part of this ritual, and to add to it any items that symbolize the gendered embodiment you're calling in. These could include photos or art, depictions of people or fictional characters, herbs or animal figures, meaningful items given to you by others, or anything else that resonates with you. I often recommend folks look to nonhuman models for gender, as human beings have done such a thorough job of perverting gender into various toxic models. For example, an acorn, as a token of an oak tree, is one of my favorite totems for masculinity. There's no wrong thing to include here. Just make it tailored to your individual self.

Ingredients

- whatever gender-affirming clothing, accessories, or prosthetics you want to bless
- any items that represent the gendered embodiment that you're calling in
- other altar items you want to add (see chapter 5 for more about altar building)
- rose water and a small bowl or cup to put it in
    - Rose water can often be found at grocery stores for culinary purposes. That's totally fine for this ritual! You can also look at a botanica or apothecary or order it online. If rose water isn't available to you, or if you want to use something else, choose anything that is safe to sprinkle on your items.

Method

1. Start by cleansing, grounding, and making some energetic boundaries around the space where you're conducting the ritual. Do what you need to be fully present with this ritual.
2. Lay the items you're blessing before you. Pause and take some time to gaze at or place your hands on each item in turn. As you engage with the item, allow yourself to connect with your reasons for obtaining the item, and your hopes and intentions for it. If emotions come up in your body or heart, let them be. You don't have to do anything in particular, but if it feels helpful, try to breathe with them and allow them to be present. You

can also take time to move intuitively, dance, shake out your arms and legs, make some noise, sing—anything that helps you be with the feelings. The goal is to avoid pushing them down and into your body.

3. When you're ready, take time to build the altar to your authentic embodiment. Take your time, placing each item with intention. You can even speak this out loud, inviting the spirit or energy of the item to lend itself to your altar and to affirming your gender embodiment. You can invite the transcestors to join the altar, and set a cup of fresh water for them if you'd like. Finally, pour a bit of rose water into your small bowl or cup.
4. Bring each item before the altar one at a time. For each item, repeat the following actions. First, use your fingers to sprinkle a little rose water over the item. A few drops is fine! Then, bring back into your mind the intentions and hopes for the item that you thought of earlier. Rub your hands together until you feel warmth between them, then place them on the item. As the warmth from your hands goes into the item, feel your hopes and intentions flowing out from your hands and into the item. Finally, send gratitude to the item itself, your past self for obtaining it, your future self for using it, and the transcestors.
5. Repeat this process for each item. When you're done, you can close out the ritual by thanking your past, present, and future self, the spirits of the items, and the transcestors. You can leave your altar up if you would like and if you're able to. It would be an excellent place to continue tending to your gendered embodiment by adding additional items as they present themselves (for example, your spent HRT bottles). If you don't have space, you could also put a few meaningful items from the altar in a small tin or jar, and use that as a more portable or space-saving option.

8

Rituals for Calling In a New Identity

For many folks, there is a great deal of ritual around name and identity based on sex assigned at birth and the name their parents choose for them. We may not call it ritual, but decorating a nursery or buying a new wardrobe for an infant certainly fulfills the criteria. People who are pregnant often get the ritual of baby showers and gift registries, and of course, birth itself is a profound ritual, with blood and pain and joy and grief often rolled into one massive threshold moment.

When we rebirth ourselves into new identities, understandings, and self-concepts, there is no prescribed fanfare. There is no massive culturally recognized moment on the scale of labor and delivery, nor multiple industries dedicated to the recognition and celebration of our process. If gender exploration and transition are depicted in media at all, it is rare and often a dramatic misrepresentation. While transitioning under the radar can seem preferable for those of us who feel awful about being perceived, I also have many clients reflect that the lack of fanfare from their loved ones can begin to make them feel invisible or even uncared for. They are changing on every possible level in such dramatic ways, but that isn't reflected back to them at all by those they trust or the culture at large.

The gender exploration process can be a slow and gradual unfolding, or a sudden awareness like a thunderclap. The journey is not linear or predictable, and while there may be similar beats for different folks, the details and specificities make a "one size fits all" approach impossible. Many folks' loved ones want to be supportive by accepting them and not making a big deal out of their transition, but transition is a big deal. And while you may not want all your people to freak out, some acknowledgment and celebration may certainly be in order. Many of the familial celebrations and rituals of gender happen around our birth. Yet no matter what age you are when you transition, you're no longer an infant: You speak and think for yourself; you have agency and opinions and emotional baggage and spiritual specificity. A new type of ritual is in order.

The following rituals are meant to provide some scaffolding and support for the process of claiming your new identity. The first two are variations on a theme—different versions of a ritual to release a name or pronouns that no longer suit you, either solo or in community, with an officiant or without one. The next ritual is based on the idea of cord cutting, which is often used in magic work, but instead of cutting a cord between yourself and another, you'll be cutting the cords between yourself and old stories you carry about yourself. We end with rituals for claiming your own gender identity, no matter how unique or expansive.

May these rituals support you fully, and create the space needed to honor and celebrate your transition!

Name and Pronoun Release and Claiming Ritual

This version of this ritual is meant to be done alone, and is perfect for those who wish to honor their process of changing a name or pronoun, but without the energy, input, or presence of others. The energy of this ritual is meant to be very gentle, but the ritual itself often feels deeply profound and impactful. Which is to say, when you read through the steps, don't underestimate how intense this ritual will feel, or how exhausted you might be afterward. Ideally, you will schedule this ritual when you have time and space to rest after. If you find yourself energized or excited after, that's awesome! Just try to stay open to what your body needs.

As always, read through the ritual and edit it as needed for your specific circumstances.

Ingredients

- items for an altar (see chapter 5)
    - some options: a cup of water, a candle, something from the world around you (rocks, sticks, dirt, leaves, flowers), feathers, a photo of a younger self or child self, transcestor photos, jewelry, any item of personal importance, things that represent the "you" that you are shedding, things that represent the "you" that you are calling in
- bowl
- paper and something to write with
- water
- optional: chalk and a rock, large enough to write several words on but small enough to pick up and fit in your bowl
- optional: journal and pen

Method

1. Do the energetic hygiene that makes sense to you. Some people may want to do this ritual either before or after taking a ritual bath. Others may only have time and capacity for a shorter cleansing process, which is fine! Just make sure that whatever you choose, you plan it intentionally. For example, if you're going to do a full ritual bath before the ritual, I would

recommend setting all of your ingredients out beforehand. If you're doing a bath after the ritual, set up your bathing space.

2. I recommend taking the time to call in your transcestors, guides, and any animal or plant spirits that you would like to witness you in this process. Just because you're doing a solo ritual does not mean than you can't have or don't need witnesses! Call out to whomever you want to be present, and welcome them in.
3. When you're ready, go ahead and build your altar. Take your time and place each item intentionally, speaking its purpose as you place it, even if that purpose is as simple as, "It feels right for you to be here." If you're using a candle, you can light it now. Gaze at the candle flame, or at the altar, if not using a candle, and set an intention to love and honor every version of yourself. After all, every version of yourself got you to this point! And that is worth honoring.
4. Take some time to identify the name, pronoun, and identity words that you're releasing. This might be as simple as shifting away from specific relationship words (mother, son, daughter, aunt, uncle, dad, sister, brother), or you may be changing everything—name, pronoun, identity words, relationship words—all at once. Whatever it is, call all of those words into your mind. Take some time to reflect on all of the parts of life that those words have been with you for. Try to bring up the positive and neutral experiences specifically, and allow yourself to appreciate them for what they are.
5. When you feel that you've fully appreciated the experience of those names, pronouns, and other words, take the time to write the words and names you're releasing on the paper, or use

chalk to write them on a rock. If you're using paper, write small if possible. When you've written them down, set the paper or rock aside.

6. Now take some time to contemplate the names, pronouns, and identity words you're calling in. Write down any and all relevant words on a clean piece of paper. This time, it can be as big as you want. Once all are written down, allow yourself to sit with these words. Contemplate what values you want to embody, what dreams you want to call into reality, and everything that you want to call in with these new words. Write down as many intentions, wishes, and dreams for your next iteration as you can think of.
7. When you're complete with that, bring the water close to you. Rip the part of the paper with the old words or name until you've separated the small part with just the words from the rest of the paper. Or, if you're using a rock, hold it in your hand now. With either paper or rock in hand, say the following words:
8. "By my own will, I release that which no longer serves me. This [name, pronoun, identity] used to reflect me, but it no longer does. With deepest love and compassion to all of the past, current, and future versions of myself, I release [say the pronouns, names, or words]. As I release this, so too am I released from it. According to my will and with the support of the transcestors and the benevolent spirits who join me in this space, I release you."
9. Place the paper or rock in the bowl of water before you. If using the rock, use your fingers to wash the chalk from it. If using paper, rub the paper under the water until it turns to

pulp. If you like, you can say, "As this water washes away these words, so they are washed away from me."

10. If it feels good, close your eyes and imagine the paper or chalk floating out into a river, stream, or ocean and dissolving into the water. Imagine the energy of it spreading and diffusing through the paper or chalk. Whatever was written here has been honored and released—it no longer has any hold over you.
11. Now, take the paper with the new name, pronouns, or words on it. Hold it to your head, and then to your heart. When you're ready, you can say this (or something similar): "By my own will, I call in this [name, pronoun, identity]. These are the words I have chosen to reflect me. With these words, I speak aloud my intentions. [Here, read your intentions or just speak from the heart.] According to my will and with the support of my transcestors, I claim all of this and better."
12. Close your eyes or let your gaze become soft and allow yourself to soak in these names and words, these intentions and dreams. When you feel ready, do something to move the energy of the work. You can put on a favorite song and dance it out, move intuitively, sing, or anything else that feels right!
13. When you're done moving, take the water and flush it down the toilet. Move into the cleansing bath if you chose to do it after, or at least wash your hands and do a cleansing breath. Take your time and move slowly with your day or evening after this. It is, after all, a massive ritual. Be gentle with yourself, and be proud. Well done!

Name and Pronoun Release and Claiming Ritual in Community

Choosing a new name and pronoun can often be a strange process, because although it can feel intensely private, the end result tends to be quite public! In order to have folks begin using our names and pronouns, we must tell them what to call us. In other words, at some point, name and pronoun claiming becomes a community act, no matter how private it feels.

People handle this type of disclosure in every way imaginable, and there's really no wrong way to navigate it, as long as the way you're using it works for you. That said, there are distinct advantages to holding ritual with your supportive friends, family, and community. When folks are invited to publicly support you, to publicly witness you, it's very different from the more private one-on-one process that many of us use to disclose our chosen name and pronouns. The public nature of that witnessing—in other words, the very fact that it is done in the context of a group—helps to solidify your name and pronouns in your community, while also allowing you to receive a more robust level of affirmation and support than you would one on one. It's mutually reinforcing.

This ritual is designed to facilitate that exact experience. Your community is declaring that they love you and that they will refer to you correctly. They are declaring this both to one another and to you! This is a powerful process and it benefits from a little bit of prework. I highly recommend reading through the previous solo version of the ritual first to get a feel for the process, then come back and read this one. Make sure to read all the way through the ritual before you attempt it.

Ingredients

- items for an altar (see chapter 5)
    - some options: a cup of water, a candle, something from the world around you (rocks, sticks, dirt, leaves, flowers), feathers, a photo of a younger self or child self, transcestor photos, jewelry, any item of personal importance, things that represent the "you" that you are shedding, things that represent the "you" that you are calling in
    - If desired, you can invite ritual participants to add their own items to the altar, and it can be a communal altar.
- bowl
- paper and something to write with
- water
- small pieces of paper and writing utensils for guests
- optional: chalk and a rock, large enough to write several words on but small enough to pick up and fit in your bowl

Method

1. **Prework**: In this ritual, each of your guests will have an opportunity to speak a statement to you. Everyone can author their own statements, but I recommend they follow this template: "I've always loved/appreciated [something about you], and I'm looking forward to seeing how you, [chosen name], carry that into the future."

   Example 1: "I've always loved how you're able to hold your loved ones with care and grace, and I'm looking forward to seeing how you, Eli, carry that into your future."

Example 2: "I've always appreciated how he shows up for me when I'm struggling, and I'm looking forward to witnessing how Eli carries this into his future."

2. I recommend giving this template to your guests in your invitation to the ritual, along with instructions about what to bring if you're doing a communal altar. When I've officiated this ritual, we've also created a "program" for the day that can be printed and put the instructions in there as well.
3. Next, you will need to decide the general elements of the ritual that you want to include. **I highly recommend having an officiant for this ritual.** It is a lot to manage all your people's feelings and questions, and hold the space, and do the ritual itself! I don't want to say it is impossible, but it would certainly be a lot to process. Your officiant can be a friend, a loved one, a ritualist, a spiritual leader—whomever works best for you. If you can, four to six weeks before the ritual, sit down with your officiant and work out the elements of this ritual. A flow that works well for me is the following:
    a. Introduction, welcome, and land acknowledgment.
    b. Some sort of marking that the ritual is beginning; I like to ring a bell.
    c. Officiant says a few words about you and your journey to this point. You can write this part together!
    d. Build the altar—either yourself, or bring your community in to add their items.
    e. If your friends and beloveds want to contribute, it can be lovely to have folks read a poem, sing a song, or some other offering to the ritual space.

f. Officiant says a few words about transformation and becoming and invites you to write down the name or pronouns you're releasing, either on paper or on the rock.

g. Folks say their prepared statements aloud, one by one.

h. You either wash the rock in the bowl of water, or you rip up the paper and place it in the bowl of water.

i. Everyone speaks their commitments (examples below).

j. Officiant gives a closing statement and ends the ceremony.

4. As you can see, there are a lot of parts to this ritual, which is normal for a community ritual. There's a lot of room to customize, and when I do this ritual with clients, we work closely to customize it into something that works for everyone. I write the ritual and then we collaborate closely on the wording and the specifics. I will offer some of my examples, but you and your officiant can decide on exactly how you want to word things, what elements to add, what elements to release, and what to change.
5. What follows is, more than anything, an example of the ritual for you and your officiant to riff off of. Do whatever you need to do to make it your own.

---

6. **Welcome**: Hello and welcome, everyone! We are gathered here to honor [name] with our presence, time, and attention. Please take a moment to silence your phones and put them out of sight. Optional: [Designated photographer] will be taking images throughout the ceremony so that the rest of us can be fully present in this space together.

7. **Introduction and land acknowledgment**: The officiant introduces themself and can say a few words about how they know you. They can then conduct some sort of land acknowledgment, if that is within your practice. Land acknowledgments may or may not be appropriate, so I will leave that to each reader to determine for themself.
8. I recommend inviting the audience to take a few deep breaths together. There is something very powerful in syncing the breath. Sometimes I also invite everyone to breathe in, breathe out, and then clap together three times. These types of somatic cues allow our bodies to become present in a space, and allow us to focus on what we're doing together—something we often need to be reminded how to do in these times.
9. **Opening the ceremony**: This is where I would ring my bell three times. The officiant can also declare that the ceremony is beginning, or some other thing can be done to your preference. The officiant should say some words here. I like to make my words about the individual's specific and unique place in this world and in their gender journey. Each person's version of this will be different, and if the officiant is not comfortable writing this piece by themself, you can write it together.
10. **Building the altar**: At this point, you can build the altar. If you're inviting others to add to the altar, I recommend you placing your items first, and perhaps lighting a candle, then inviting folks to come and place their items. This part might benefit from some background music, which you can play on a speaker, or if one of your friends is inclined to play an instrument or sing, you can do that too.
11. **Words on transformation plus writing the name or pronouns to be released**: The officiant can say some words here about

the nature of transformation, a bit about the transcestors and the importance of them, or whatever else feels important. Here's an excerpt of one of my ceremonies:

"This ceremony is a recognition of one facet of a process that is familiar to all living things; the continuous spiral of life, death, and rebirth. In spring, a leaf grows on a tree. In summer, the leaf converts sunlight to food and gives the tree what it needs to survive. In fall, the leaf changes and falls to the forest floor. Once fallen, it becomes food for other organisms in the environment. Over time, it is reduced to the chemicals that the soil needs to sustain many things. Each of these organisms takes up that nutrition, and grows stronger for it. In this way, nothing is wasted, and the cycle continues.

"The transitional process we acknowledge today is like this. Nothing of this life is wasted. We celebrate all that has been composted and will be brought forward into [name's] life. We celebrate the fullness, the depth, and the breadth of their humanity, the beauty, complexity, pain, and transcendence of their experience. And while we celebrate this moment and this milestone, it must be remembered that this is not the end of their journey. Releasing a name is a powerful act, but we do not gather here to mark a "happily ever after." This is a beginning, like many new beginnings they have experienced in their life so far, and the many new beginnings that will come after today."

12. After the officiant says their piece, they can direct you to write the name or pronoun that you're releasing onto the paper or the rock.
13. **Community participation**: At this point, folks are invited to share their statements. As an officiant, I like to go first to model it for everyone and to break the ice. It's up to you if you want

people to go in order or move when their heart tells them to. The officiant should allow time and space for people to work up their courage if you're using the latter method, though, and should ask a few times if everyone has spoken who wants to.

14. **Releasing the name or pronoun**: Once everyone has said their bit, the officiant can direct you to wash the rock in the water, or rip the paper and place it into the water, rinsing it around with your hands until it gets mushy and dissolves. Take your time with this step.
15. **Community commitments**: I like the crowd to speak two commitments together, though you can do whatever number you like. It is important that these commitments are spoken with your new name. You can put these commitments in the program so that folks know what to say, or the officiant can have everyone repeat after them. Here are the commitments I like to work with.

    [Name], we witness your multitudes and honor your path.

    We hold and support you as we embrace your new name, [Name].
16. **Your statement**: At this point, you can say any words you want, even just "Thank you" to everyone who is supporting you and who chose to be there. This is your ritual, so feel free to make this part your own.
17. **Closing out**: The officiant can close out the space with gratitude to those who joined, to the transcestors, to the land and the web of life, and to you. Everyone can breathe together one last time, and then the officiant can ring the bell. It is also fun to invite everyone to make a joyful noise together as the bell rings, should your space allow for it.

18. **Energy release**: I recommend having some sort of energetic release as folks transition to the next activity. A potluck is great for the post-ceremony time. It's wonderful to play some music, let people dance it out a bit and give hugs. Whatever feels good to you and your officiant is perfect here.
19. And that's it! Regarding the altar: If it is in a stable space, you can leave it up during the rest of the festivities, but be clear with what folks should do with their additions. It's totally okay to tell people to take their additions home with them, but if you're doing that, I would reiterate it in the program and have the officiant remind everyone at the end of the ceremony. If you are keeping the items people bring, make that clear as well, and know in advance what you intend to do with them. Any perishables on the altar can be eaten or composted according to your preferences and cultural practices. The water from the bowl can be poured into the earth or flushed down the toilet if it has paper in it. If you're doing this, be mindful of the type of paper you use, how much of it, and how robust your plumbing is.

Ritual for Releasing Your Attachment to Outdated Stories About Yourself

It is very common that we carry a variety of outdated stories about ourselves. Sometimes these stories are relatively harmless and just need a gentle update. For example, let's say that you used to hate olives when you were a child, but now you enjoy them. That's a pretty simple update, and probably doesn't carry a lot of emotional charge. Other times your stories can be quite damaging and detrimental and can even impact how you show up in the world and what decisions you feel are available to you.

When I first started T, I had a story that I couldn't possibly claim a male identity. At that point I had identified as nonbinary/masculine for years, but always shied away from saying I was male or even implying it. If I thought about claiming a male identity, I felt a massive cringe in my body, because my story was that I was a liar, simply trying to appropriate an identity that didn't belong to me. My story was that I was just an annoying person who wanted all the benefits of being a man but who categorically was not. I wanted to take T because everything in my body was screaming for it, but that story was powerful, and even as I went in to the doctor for my prescription, I was terrified.

It wasn't until I reckoned with the fact that I deeply loved everything happening in my body as a result of T that I was able to embrace my maleness, and it wasn't until I embraced my male identity and affirmed my masculinity that I was able to lean more fully into a more gender-expansive identity and aesthetic, but that's a story for another time. The point is, the stories that you have in your body don't have to be negative, but when they are, they can be very powerful and very controlling.

In the tarot, this dynamic would be associated with the Devil card from the major arcana. The Devil is, even in most gentle decks, a frightening image. There's usually a big scary monster and some people in chains, or some variation of that imagery. The funny thing about the Devil is that the archetype often reflects the ways that we put ourselves into what I like to call "un-sexy bondage." We do this by constraining ourselves unnecessarily with stories about how we are bad, or inadequate, or unlovable, or any number of painful things. These stories are not true, but they are powerful, and as long as they control us, we have very little access to our power.

But the story of the Devil doesn't end there. Often the most frightening thing we can do is truly face those stories and dismantle them. If you've been in chains long enough, freedom starts to feel like danger, and facing that head-on starts to feel like a fool's errand. In the same way that the chains are actually stories that we hold about ourselves, the monster is the inflated danger we've convinced ourselves of. It is the worst-case scenario, the anxiety spiral, the unfiltered fear of rejection. When we steel ourselves and face that fear, it often evaporates before us, allowing us to experience true liberation.

Does this sound familiar? Do you know what your stories are? If you're struggling to identify them, start paying close attention to your actions and your reasons for those actions. Do you stop yourself from going for opportunities that you really want? Are you telling yourself that you aren't trans enough? Does your imposter syndrome pop up in other areas of your life?

This ritual is one that you can start right now, but you may need to repeat again and again. I've done rituals like this over and over again, addressing various nuances and depths of a specific challenge. For particularly sticky stories, I've done an entire lunar cycle of ritual like this. For others, it's been one-and-done. For challenging topics, I have come back to this ritual in various iterations and varieties again and again for literal years. Only you will know what you need.

Remember that you can reread the earlier chapters for ideas about timing, repetition, and other logistics. Read through the ritual as it is written, then adjust it to your preferences and needs. And remember, you are empowered to liberate yourself from these stories! You don't need permission. You've got this.

Ingredients

- 2 chime candles: black is best, you can also use red or white, or if there's a culturally specific color you prefer, use that.
- candle holders for the chime candles: I usually just use tinfoil to hold them up. If you do this, make sure your tinfoil candle

holders are on a plate as they are liable to spill wax everywhere as they melt.

- thumbtack, toothpick, ballpoint pen, or candle scribe: basically anything with a pointy end that can be used to scratch something into a candle
- lighter or matches
- heat-proof dish and something to set it on
- string: natural fiber string is best, not too thick
- whatever you need to cleanse your space and body (see chapter 4)
- paper and something to write with
- scissors

Method

1. Before you begin the ritual (this could be earlier the same day but could also be in the week or two leading up to the ritual), take time to really get clear on what you're cutting cords to. Don't do everything at once! Choose one dynamic that no longer serves you, one outdated story, one piece of identity that no longer resonates. Whatever it is, choose only one. Take time to write down, or at least think on, all of the feelings and actions associated with that thing. Take your time with this, and really let yourself feel into the fullness of how this moves you. Then see if you can put one or two words to encompass this thing. The words are not meant to get into every nook and cranny of what you're cutting cords to. They might be something like "self-hate" or "not enough" or "I'm bad." If words feel challenging, you can make a simple symbol that represents what you're cutting cords to instead.

2. When you've chosen your word or symbol, do as robust a cleanse as you have time for. Cleaning your house would be a great idea, but you can also clean your room if that's more accessible. No matter what else, cleanse yourself as robustly as you can. Be mindful of your capacity, and do your best.
3. When you're ready, set up your ritual container according to the instructions in chapter 5 under "Steps to Opening a Ritual Container." Call in your guides, protectors, transcestors—any benevolent and supportive forces that wish for your highest and best outcomes. This ritual is like surgery for your energy. The more support that you have, the better.
4. Next, take your candles. In one candle, carve your name. If you want to, you can also carve your birth date and the city you were born in. In the other candle, carve your words or symbol. If you want, you can anoint the candles with an oil, such as an uncrossing oil. If you don't have any appropriate oils, don't worry about it.
5. Tie the string onto each candle such that they are connected. Take your time and consider your connection to this dynamic or story that you're releasing. Consider whether this story has provided you with protection or support in any way. Let yourself be grateful for the work it did to try to protect you. Let yourself feel any other feelings that arise: rage, grief, gratitude, pain. Take your paper, and on one side, write your name. Draw a line down the center, and on the other side, write what you are releasing. Place the paper between the two candles so that the side with your name is closest to the candle with your name.
6. Light the candles.
7. Now take the time to consider all the ways that this connection to this story or dynamic or outdated version of you has caused discomfort or harm. Again, let yourself feel the emotions that

arise, acknowledge them, but try not to lose yourself in them. When you're ready, consider what good could come if you were free of this story. Let yourself imagine at least a few possibilities. When you have a strong idea in your body and heart of how good it might be not to have this story any longer, take your scissors and carefully cut through the paper and then the string. As you do this, speak these words (or something similar): "By my free will, and with harm to none, I sever the cord that binds me to [the things you're releasing]. Though you were once a source of protection, you are no longer. I care for myself, and therefore I release you with care. As I release you, I too am released. According to my will and with the support of [my guides, transcestors, etc.], so may it be!"

8. Take the half of the paper and place it in a heat-proof dish. If you can safely do so, light it on fire and let it burn to ash. You can say, "As this smoke rises and drifts away, so too does my connection to you. So may it be." Wait until the paper is done burning. If you can still feel that dynamic or relationship in your body, use the scissors and snip the air in front of the part of your body where you feel it. If you feel it in multiple areas, snip in front of multiple areas. Be mindful not to accidentally snip yourself!
9. When you're done, extinguish the candle that represents what you've released. Leave your own candle burning until it has burned fully out (don't leave it unattended). Bury the candle of what you're releasing with the ashes of the paper. If you can't bury things, dispose of them in some other way, preferably in a public trash can, not too close to your house.
10. Take time to cleanse again after the candle goes out and after you've disposed of the other candle and the ashes. You can use any of the methods in chapter 4, but I recommend a full ritual bath if you can. If not, just do the most robust cleanse you can.

11. It is important after this ritual to be very well hydrated. Try to do daily energetic hygiene, even if just for a few minutes. If you can, keep track of how you're feeling in your journal. Every month or so, take time to look back through your journal to see if you're still dealing with the same dynamics, or if you need to do a more targeted version of the ritual. And remember, you can do this as many times as you need to!
12. Pro tip: It can be helpful to do this ritual during the waning moon—that is, after the moon has been full and before it is new. This is a time of release, so it lines up with the intentions of the ritual.

Rituals to Claim Your Unique Gender

Whether you want to claim or call in more masculinity, more femininity, more androgyny, more genderfuckery, or more of the unique, uncategorizable gender energy that makes you, *you*, a ritual can serve you. This ritual does require some prework, though, as it's important for you to be clear on exactly what energy you're calling in. Even if you're comfortable with a binary identity, what I mean when I say "masculine" and what you mean when you say the same thing may indeed be very different things. My masculinity is specific—it is well-cultivated, animistic, based in the things that resonate and light me up, specifically. Your masculinity or femininity may be something very different from how those things are usually interpreted in modern culture.

Then, of course, there is the gender that exists in a more liminal space, outside the binary, despite the binary, without regard to the binary. A.k.a. the rest of the gender universe. The same general formula for a ritual can be adapted in any way that resonates to your specific gender embodiment. Your clarity on exactly what gender you're claiming does not need to be specific in terms of language or legibility to anyone outside yourself. If you know what you're claiming, whether that's a vibe, a label, an energy, or some other gender thing entirely, you're good to go.

So start by identifying for yourself what it is that you're claiming. Then, go ahead and take the time to create an archive of correspondences. Roughly speaking, you're gathering items that you can add to an altar. However, your altar in this case could be an artwork, a collection of poems, a collage, even an herbal tea or smoke blend. Or it could include those things as well as more traditional altar items (see chapter 5). It may take some time to identify these items; that's okay. Again, clarity is key, so if it takes time, trust the process.

You can employ quite a lot of intuitive work in this process. Go to estate sales or thrift stores or flea markets or a forest and wander around. See if you feel drawn in any particular way to any particular item. It probably won't feel like angels parting the heavens and shining a light on the exact thing you need, although it might; it may be just a gentle desire to go this way or that. Let your gut guide you, feel the resonance with your energy, and move accordingly.

As you gather this aligned gender archive, you will also want to gather things that represent the genders or gendered energy that you're releasing. This doesn't have to be things that you hate. Instead, focus on anything that feels like it belongs to another version of you, anything you've energetically or literally outgrown. Things that perhaps mattered a great deal to you but that no longer feel like they are for you. Make sure that anything you add to this category is something you're ready to both energetically and physically part with. There's nothing wrong with keeping some items as keepsakes or reminders of who or where you've been, but this ritual involves getting rid of the things that you've outgrown, so it's important to be genuinely ready to do so.

If you're unsure about your energy or what you're claiming, go back to the ritual exercises in chapter 6. When you're clear on what you're claiming, and you've gathered the items that correspond to it, you're ready for this ritual.

Ingredients

- anything that you need for cleansing, grounding, and boundaries (see chapter 4)
- cutting tool: a plain old kitchen knife, a ritual scraping tool, an athame, a piece of selenite, or anything else that you want to work with for the purpose of cutting or scraping energetic bonds
- all the items that you're releasing
- bags, boxes, or receptacles of some kind to hold what you're releasing
- blanket or piece of fabric
- paper and something to write with
- your most gender-affirming outfit plus all necessary accessories and prosthetics
- all the affirming items, archive, art, etc. that you gathered, per the introduction
- optional: journal

Method

1. Before you begin, make sure to gather all the items needed for the ritual, especially the altar items and archive mentioned above.

2. Start with a robust cleanse. If possible, do some form of the cleansing bath from chapter 4. As you cleanse, specifically cleanse away anything that isn't aligned with the gender you're claiming and calling in. In a very real way, this ritual is your statement to the universe, to the web of all life, of who you are and what you're available for and not available for. If you want to add to this cleanse, you can begin it with a gender-affirming haircut, shave, or trim (see the rituals in chapter 7). Regardless, you want to be as cleansed as possible on as many different levels as possible. Ground your energy and take the time to boundary and protect your space as described in chapter 4.
3. Put on your most gender-affirming outfit. If this requires accessories, makeup, prosthetics, or anything else, go for it. If this requires specific types of grooming, do them. Take your time and really set yourself to feel deeply grounded in your gendered energy, to feel as *you* as you possibly can.
4. Begin by gathering the things that you're releasing into a box, bag, or some other receptacle. If these things need to be sorted (some to donate, some to trash), you might want to have multiple receptacles. As you gather the items and sort them, take time to record on your paper what energies you associate the items with, and what you're releasing. If the items have strong associations or representations, take the time to reflect on those associations, send gratitude to your younger self, and gently release your need to hold onto the physical item. Remember that every version of you is always with you, but you do not need to hold onto all the energy associated with the belongings of a past you.
5. If there is gender associated with the items, address it specifically. You can write it down (for example, "This was my masculinity extinction burst!"), do some intuitive movement, breathe

it out, or anything else that feels right. Once you've sorted the items and written down the associations, you can take your cutting tool and draw it over the bag or bin where the items are included. Say these words or something similar: "I cut away the cords that bind me to these items. I release them and all that they represent. I cut away the energetic connections to these items. With gratitude, I release them, and I release the version of me that they represent."

6. Take the paper with all the associations that you're releasing, and rip or cut it up into as many pieces as you want. Scatter the pieces over the bags or boxes of items, then put a black or dark blanket over them. They are now released and cut away.
7. Do a quick cleanse with breath, hydrosol, or smoke. Your energy is now fresh, cleansed, and open to that which you're claiming. If you need to ground and boundary again, this is the time to do it!
8. Next, bring in the items that represent what you're calling in. You can place them on an altar if you want, or in some way arrange them such that you can look at them. Take some time to breathe, to feel into the energy of those items and into all that you felt while you were gathering them. If it resonates to you, write in your journal or on a piece of paper all that you are calling in. If there are words for your gender, write them here. If there are shapes, colors, or other visual indicators, draw them or record them. If there are movements, dance them.
9. When you've recorded what you're claiming in whatever way you need to record it, close your eyes, soften your gaze, or focus on one main point. Take some time to pull that energy deep into your body, and let it move you. As best you can, embody the energy that you're calling in. You might shift

your posture, squaring your shoulders or cocking your hip. You might make your body smaller or larger; you might cry or dance. Whatever you do, focus on making the shape of the energy you're claiming with the body you have right now.

10. You can take time to move around in this shape. You can speak or sing or just walk around your room or the space you're in. If you want to, you can even walk around the block in this energy, really feeling and projecting it to the greatest degree that you can. When you feel you've fully embodied the energy as much as is possible for this moment, come back to your altar or archive. Sit, stand, or lay before it. Speak words of power and blessing over yourself. You can use these words if you want, or write your own.

    Benevolent transcestors,
    Those who know what it is
    To shift embodiment
    To move fluidly from one to another
    To move to the beat of your own heart
    To move to the rhythms of your own body,
    I call upon you.

    Bless this body,
    Bless this embodiment,
    Witness that which I am claiming
    Witness that which I am calling in.
    [Here, you can name what you are calling in.]
    With ease, with love, with gratitude,
    Push all that I have released far from me
    Let it be recycled into the great web of all things
    Let it be repurposed where it is most needed.

    With ease, with love, with gratitude,
    Let the energy of my selfhood flow to me and through me

Let it bolster me and protect me
Let it attract to me all that serves me
Let it repel all that would harm me.

I claim this energy now,
Energy of [name what it is].
This is mine, and it can not be separated from me.
May it serve my highest and best purpose
May it serve my community
May it serve the web of all life.

May it be, may it be, may it be.

11. At this point, you can journal any thoughts, feelings, or realizations that are coming up for you, or you can dance it out, move intuitively, or anything else that feels like it solidifies your work in this ritual. When you're done, send gratitude to the transcestors, to yourself, to your helpers and guides. You can leave your altar or archive up and continue to tend it, or you can take it down. For the items under the blanket, dispose of them as necessary. For the paper pieces, throw them away in a trash can that is not inside your house.

9

Rituals for Misgendering and Exposure to Transphobia

Introduction

Changing the name or pronoun folks use for you is often a powerful part of transition. While everyone's journey is unique, this more outwardly visible step often reflects a great deal of deep interior work. Claiming a new name or pronoun can conjure many feelings, including relief, comfort, euphoria, trepidation, tenderness—you name it. Hearing folks use your name and pronoun correctly can be a thrill, an affirmation of your identity, and also a signal of how much that person loves and understands you, or is willing to try.

When folks do not use the correct pronouns or name for any reason, this is known as misgendering. For many, misgendering is a painful part of trans life. Not everyone feels dysphoric when they are misgendered, or when the wrong name is used, but it can highlight that the person you're speaking to doesn't see you, not really. With or without dysphoria, this can trigger any number of negative feelings. If you're misgendered throughout a whole day, or week or month or life, these negative feelings can add up to create a deep well of pain. Death by a thousand cuts, as the saying goes.

I find that misgendering is particularly prevalent for clients who use gender-neutral or neopronouns. Many English-speaking folks we meet in the context of our daily activities will default to a binary he or she pronoun. The prevalence of this social norm means that folks using they/them or any of the neopronouns are consistently and constantly misgendered, regardless of their presentation.

For those of us who are privileged to avoid misgendering, transphobia is still a constant feature of the water we swim in. We are exposed to it in myriad ways in social interactions, entertainment, and social media. As I write this in 2025, the transphobia expressed by political leaders and parties is at an absolute fever pitch and shows no sign of slowing. Many trans people that I talk to are experiencing this as a sort of constant drain on their energy. They feel agitated, exhausted, and worn out on a bone-deep level.

The accumulation of transphobia and the fear and hatred aimed squarely at trans folks right now is a toxic sludge with a deep impact on our well-being. More than ever before, I have heard folks internalizing the transphobia around them in deep and painful ways. This creates a profoundly negative feedback loop wherein it can become incredibly challenging or impossible to access your own trans energy or the joy, contentment, and peace that transition can bring. It's a crisis on multiple levels, and each and every one of us needs the proper tools to navigate it as best we can.

I hope that you, reader, are in a different context, but I know that transphobia is a feature of modern Western culture and not a bug. Whether you're on the front lines of the transphobic legal battles or just a trans person who is trying to get by, these rituals will help to avoid that low-grade or high-grade inflamed feeling of constant exposure to transphobia.

The first ritual is focused on protection and shielding for the times you are misgendered or exposed to transphobia, whether it is a small annoyance or a full-body inflammation. This ritual builds on the shielding work discussed in chapter 4 to create robust energetic protection. It also calls on the energy and care of the transcestors in order to facilitate shielding and protection. While the focus is on misgendering or other forms of transphobia, this ritual can be modified and used for any situation where protection and shielding are needed.

The second ritual will help your body return to baseline and your nervous system realize that you're no longer in danger after a challenging, painful, or difficult experience involving transphobia, misgendering, or any other personal discrimination, violence, or aggression. This ritual is functionally for cleansing, and can be used with or without the previous ritual.

I recommend using the protection and shielding ritual before a time when you know that you will be exposed to transphobia, such as a protest, at a transphobic workplace, or at a family gathering. The cleansing ritual should be used after you've been exposed, or as a way of getting the energetic gunk off so you can reengage your energy from a fresh place.

Ritual for Protection and Shielding Against Misgendering and Transphobia

This ritual is highly customizable. The level and intensity of protection that you will need may be different for different circumstances—for example, a family gathering and a doctor visit could both be harrowing, but the circumstances are different enough that you may wish to customize the ritual to each. If you need a refresher on protection work, go to the shielding section in chapter 4.

Ingredients

For this ritual, the supplies are entirely optional. Read through the ritual first, then assess what is available to you and what feels most helpful.

- protective herbs
    - in whole or cut form, fresh or dry, tincture, essence, or hydrosol
    - Some protective herbs I like to work with are yarrow, rosemary, clove, basil, angelica, hawthorn (especially the thorns), rose (especially the thorns).

- protective adornments
  - a broad category: amulets, talismans, jewelry, clothing, makeup, gender-affirming prosthetics or clothing, anything that you can wear on your body that provides you with a sense of protection, safety, or buffer from the world
- any gender-affirming clothing or accessories that you want to wear into the situation or hold in your pocket
  - There may be overlap with the protective adornments, but for this category I mean any clothing or accessory that makes you feel good and affirmed in your gender, regardless of whether or not it helps you feel protected.

Method

1. You can conduct this ritual anywhere, though it may be helpful to have a bit of quiet and privacy if that's available to you. If it's not, you may use noise-canceling headphones to create your own little bubble (this is something I do often). Mainly, though, this ritual is written to be accessible almost anywhere because we do not always have time and space for a large ritual when protection is needed.
2. Start with a quick cleanse. You can use the breath cleanse from chapter 4, or wash your hands with soap and salt. You could also imagine a cleansing light moving down your body from crown to toes.
3. Do the personal bubble shielding exercise ("A Basic Shielding Practice" in chapter 4). Put briefly, root into at least one element. Pull that energy into your body. Let it mingle with your own energy. Use your breath to push that energy out, as if you're blowing a bubble, until it surrounds you on all sides. Top to bottom, side to side, front to back.

4. Take the time to call out to your transcestors, either out loud or in your mind. If you have known transcestors, you can take this time to call to them. If not, call to those who can recognize your current need. For example, "I call to all my benevolent transcestors who know the pain of misgendering." Or, "I call to all my benevolent transcestors who can shield me from transphobic violence." If it helps, you can imagine their presence as a specific color, texture, image, or smell that you can add to your bubble.
5. Next, call any other protectors, guides, and helpers to join, fortify, and strengthen your shield. If you're using protective herbs, this is a great time to incorporate them. How you incorporate them depends on what you're using. Make sure that if you're ingesting anything, it is safe to ingest.
6. I like to call in thorns to create a thicket around me. Thorns are excellent protectors and are boundary experts. Your thorns can be from hawthorn, rose, cactus, or even porcupine—whatever thorny friends feel supportive and available to you. This can also be a great moment to connect to bedrock (for rooting deeper) and archetypes, animal friends, plant friends, or other beings who bring you strength.
7. If you work with adornments for protection, including specific amulets, jewelry, or clothing items, take the time to ritually add them to your body. As you adorn yourself with these items, ask them to protect and hold you, to assist you in holding your boundaries, or anything else that feels protective and supportive. You can also choose specific clothing items because they make you feel affirmed and good in your body. If you need to be stealthy, this could be as subtle as a wristwatch or something that you can carry in your pocket. Think of each of these adornments as layers that you're adding to your shields, bits

of armor. Each of these can deflect direct harm; they are your allies in this experience.

8. When you feel complete, take in three deep breaths with hard exhales. Thank all those who are protecting you. Move forward with your day.

Ritual for Cleansing Energy After Transphobic Experiences

This ritual calls on the cleansing skills that you learned in chapter 4, so it may be supportive to revisit the "Cleansing" section of that chapter before you proceed. The basic idea behind this ritual is simple: Misgendering and other forms of transphobia leave a negative energetic charge, or even just an icky feeling, in the body. Sometimes that ick is worse than other times, but it builds up like grease on a stovetop. The earlier that you cleanse it away, the less buildup you have to contend with later.

If you work in a transphobic environment or you're otherwise exposed to misgendering and transphobia on the daily, you may wish to cultivate a version of this ritual that is sustainable as a daily practice. That might mean doing the handwashing version of the ritual, or even using the breath instead of water. Still, I recommend you do the ritual as written a few times to help you get the hang of it, then experiment and find something that works for you.

Ingredients

- shower, bath, or some way or place to wash yourself
    - If you're in a place where you don't have access to these things, or don't have time for a full bathing session, you can

do a version of the ritual where you just wash your hands. That being said, I personally get the most from this ritual when I am able to wash my full body.

- whatever soaps and scrubs you use on your body to feel clean
  - I love using a sugar scrub (where applicable; not on my face or scalp) for this ritual. Having a scrub with charcoal, salt, or cleansing herbs is also really lovely. If you have magical or talismanic soaps that are meant for cleansing, healing, and protection, this is a great time to use them.
- optional: candle
- optional: playlist of music that makes you feel good, makes you want to dance, or makes you want to be in your body
- lotion, coconut oil, herbal oils, or whatever you use to moisturize after a shower
- comfy robe, towel, PJs, or just something to wear after the cleanse

NOTES ON THE SETTING

- This ritual is best done in a location with water: a sink, bathtub, or shower works great. Work with what is easily available to you.
- After the washing portion, you will be invited to reconnect with your body through moisturizing and gentle engagement, so it can be helpful to also have access to a bed or bedroom or a place where you can relax and tend to yourself.

Method

1. If you can, start by spending some time tending to the parts of you that hurt. Acknowledge them; let them know that you know they are hurting. You can put your hand over your heart

to provide a little pressure on your chest and let yourself know that you are there to hold space for these difficult emotions.

2. Turn on your shower and let it come to the right temperature. While it warms up, light your candle and start your music or playlist. If you work with candles and generally do candle magic, feel free to use this time to add a candle spell to this ritual.
3. Get in the shower and let the water wash over you. Then name the instances or experiences that are causing you discomfort. Notice whether you feel any sensations in your body as you name these things. If any pain, tension, tightness, or other discomfort comes into your body as you do this, spend some time breathing into that space.
4. When you feel ready, start washing your body from the top down. I start with shampoo on my hair, face soap on my face, and then a sugar scrub from the neck down. The sugar scrub isn't necessary, but the roughness of it really helps me with this ritual.
5. As you wash, visualize yourself sloughing away the misgendering and the things others have projected onto you. Being mindful not to over-scrub or damage your skin, release the painful projections and misgenderings into the soap suds. Imagine that they are sitting just on top of your skin, and are easily lifted and washed away in the lather. I recommend getting very intentional here and cleaning your body gently but thoroughly.
6. When you're ready, let the water wash over you. As the lather mixes with the water and flows down the drain, imagine it taking the misgendering, pain, and fatigue from your body with it. If it feels okay, place a hand over your heart. If not, just focus

on sending some love to the part or parts of yourself that feel safe to send love to.

7. Once you're out of the shower, dry off thoroughly and then bring out your lotion or body oil. Spend time tending to your body, rubbing lotion into your skin, and affirming yourself. For folks who are very comfortable with their bodies, this may be a time to say out loud things like "I love you" and anything else that feels affirming. If parts of your body give you dysphoria, you can pay more attention to the parts that feel safe, or simply close your eyes and thank your body for the work it does moving you through the world. You can keep your music on during this process if you want, and even spend some time dancing, singing, or moving around.
8. If you work with anointing oils, this is a great time to apply them, giving special attention to the parts of your body that tend to hold tension.
9. If it feels good, you can speak a blessing over yourself. An example blessing is included below.
10. Wrap yourself in a soft robe or towel and spend as much time as you can relaxing and resting. You can cuddle with a pet or a partner, eat a delicious meal, drink some chamomile tea, read a book, or anything else that nourishes you.

BLESSING

With deep love and gratitude, I call blessings to this body, which has endured so much.
I call in gentleness, softness, and sweetness.
I call in rest and relaxation.
With deep love and gratitude, I call in the blessings of my transcestors, known and unknown.

> May I bring joy, peace, and healing to all of my transcestors,
> and may I walk in the blessings and love of all benevolent transcestors.
> I call upon your wisdom and your love, I call upon your knowledge and your grace. I call upon the lessons that you hold.
> With deep love and gratitude, I call in blessings and protection.

10

Rituals for HRT

For so many of us, hormones are magic and miracle, devotion and spell. I didn't fully understand my transness until testosterone began creating changes in my body. Though I was terrified of starting HRT, when I noticed the hair coming in on my belly and my lip, I couldn't ignore the thrill of excitement that went through me. I couldn't accept that I wanted to be seen as a man until I had the benefit of testosterone flowing through my body in high enough concentrations to be experienced viscerally.

That same tension is something that I see in my clients all the time—desire and fear, sometimes in equal intensity. It is the fear of the unknown that stops so many people from trying hormones, or delays them. Of course, HRT isn't for everyone. Not everyone enjoys it the way I did. Not everyone benefits from creating a bespoke hormonal environment in their body. But many people do, and it is hard, perhaps impossible, to know how HRT will feel for you if you don't at least try it.

Even still, I believe that it is not really advantageous to approach your fear and trepidation as an obstacle to be bashed through. When I was terrified of hormones, I forced myself to reach out to the doctor anyway. I had a lot of questions and concerns, but I was scared to ask them or discuss them. My doctor ended up being truly lovely and incredible, but I didn't feel like I could count on that, and it took me a while to trust her. In other words, I could have used a gender doula.

While these rituals cannot be a full replacement for working with a gender doula, my hope is that they will help you work with your body and all of your complex feelings about HRT instead of trying to shove them down or fight them. In the first ritual, we work on preparing the body for HRT. This is a ritual I have recommended to many of my clients, and always with excellent results. The next two rituals are meant to be done with your dose of HRT. While doing a daily or weekly ritual with HRT may seem like a lot, these rituals can be tailored to your needs, and may be most supportive in the early part of medical transition.

The final two rituals address two impacts of HRT that are most common for my clients to need support around: feeling too much, and not feeling enough. These rituals are for those who can't stop crying, and those who can't start crying. They are for those who feel sudden waves of anger, rage, or grief, and those who feel numb and odd. HRT can change where emotions live in your body and how they present themselves. These rituals are for those moments.

Even if you've been on hormones for some time, these rituals may be helpful to you. As always, read through them, take what works, and leave the rest.

Ritual to Prepare the Body for HRT

This is a simple ritual. It is not long or ornate, and it requires no special materials or ingredients. It is also a powerful ritual, one that creates the space to shift your relationship to hormones, but also to your body itself. In this ritual, you are practicing a deep form of consent with your body and your self. Instead of forcing your way through fear, panic, or trepidation, you are gently asking your body to open itself to the experience of hormones, to allow them to do their work inside you, and to be present with you for this process.

If you haven't done the rituals in chapter 6, it would be good to read them over and to do them if you feel so inclined. These rituals lead you through building trust, connection, allyship, and love with your body. All of those skills will be helpful in this hormone ritual. However, you do not have to do those rituals before you do this one if they seem daunting or if you're not feeling open to them. This ritual is powerful on its own.

The only ingredients you need for this ritual are time, yourself, and hormones. The hormones are optional, and I will explain the ritual with and without them. If you want to start this ritual before you actually have your hormones in hand, that's awesome! The most important ingredient here is simply yourself.

Method

1. Start by doing whatever you need to do to cleanse, ground, and boundary yourself. You want to be fully settled into your own energy, here, and not distracted by anything or anyone for the duration of the ritual if possible. If carving out uninterrupted time is tough for you, you can make it a shorter and more frequent ritual. Just do your best to be fully present while you're doing the ritual itself.
2. Once you're cleansed, grounded, and boundaried, either take the hormones in your hand, or hold the idea of them in your mind. Greet your body and tell it that you're preparing to start a hormone journey. If you want to, you can explain to your body what hormones you plan to take, and what your expectations are for that experience. If you're holding the hormones in your hands, ask your body to feel their energy. Ask your body if it is willing to open itself to these hormones, and to the experience of having them in your body.

3. If your body gives you a "no" or a "not right now," honor that. Say thank you. Send gratitude to yourself for that answer, for showing up, for asking the question. And then put away the hormones and move on with your day.
4. If your body gives you a "yes" of any kind, honor that too. Say thank you and send gratitude to yourself for that answer, for showing up and asking the question. You may wish to get more than one "yes" before you physically start hormones. You may wish to do this ritual and get a yes for weeks before that yes feels load-bearing. Listen to yourself, and honor your timing.
5. Remember, this is a consent process. Some folks will need to continue this deep consent process well into their hormone journey. Some folks will need to revisit it after being on hormones for some time. Some folks will do it once and never again. There's no right or wrong way to work with your body's energy and consent. There's just the way that is best for you, and only you can know that way.
6. Above all, try to send gratitude to yourself for this process, for working with your body, for honoring your needs.

Ritual for Your First Dose of HRT

Once your body has said all the yesses that you need to hear, and you're ready to begin actually taking your HRT, a new ritual is in order. I often recommend that, for your first dose of HRT, you really give yourself a lot of time and spaciousness. It is lovely to treat this as the powerful, unique, beautiful moment that it is, even if you're not feeling that power in your body. Some folks may want to make this ritual private, while others could be inclined to include community, partners, or loved ones in this process. This ritual might also benefit from being held by a ritualist or officiant, if that feels right to you.

The goal of this ritual is to set intentions for your transition, to celebrate yourself and your miraculous body, and to embark on this exciting new part of your journey. This can be a lovely ritual to do after a cleansing bath or shower, especially if you use the bath or shower time as a way to release old or outdated versions of your self-concept, negative thoughts and feelings, and anything else that no longer serves you. I've built cleansing into the beginning of the ritual, but what form of cleansing you use is your call.

Remember that your body is preparing to shift in some pretty phenomenal ways, and this ritual is both an acknowledgment and a celebration of that fact. So have fun with it, make it yours, and do whatever you need to feel your own magic.

Ingredients

- your HRT
- whatever you need to take your HRT: syringes, needles, alcohol swabs, water
- anointing or blessing oil of some kind
- journal, loose paper, and something to write with
- fireproof dish and a lighter
    - Working with fire is wonderful but can be prohibitive based on your living situation, fire alarm situation, and fire risk situation. If fire is not available to you, you can simply tear your paper into tiny pieces and soak it in salt water before flushing it down the toilet (writing on toilet paper might be best for this use; don't mess up your plumbing), or you can put it in some salt and dump that in a dumpster somewhere that isn't your house.

- candle (a chime or taper candle is best, as you will need to have space to scratch words into it) and something to inscribe the candle with: ballpoint pen, candle scribe, anything with a point
- optional: tarot or oracle cards, or a divination system of your choosing
- optional: camera (can be a phone camera)

Method

1. Start by writing down anything and everything that you're releasing. These can be gender-related traits, but they do not have to be. Anything that you're done with, that you're ready to move on with, write it down on a piece of paper.
2. Using the cleansing instructions in chapter 4, do the most robust version of cleansing that is available to you. As you're cleansing, focus on releasing what is on the paper, and letting it go. When you're done cleansing, rip up your paper and place it in the fireproof bowl. Light it on fire (or put it in salt or salt water) and say, "I release these things and let them go. This energy no longer has a home in me. May it be scattered far from me, and transmuted into helpful energy for other beings."
3. If burning your paper, wait until it is fully ash, then pour water over the ashes and flush them down the toilet. If it is in water, flush the water. And if in salt, set it aside to dump later, but do not wait more than twenty-four hours to dump it if you can help it.
4. Now, take some time to ground and put up energetic boundaries. When you're ready, take time to dream up your intentions for HRT, or for yourself in general, as you embark on this new part of your life. Write these down in your journal. For each intention, create a keyword that relates to that intention. If sigil work is in your practice, you can use that here as well.

5. Once you feel complete with naming your intentions and giving them keywords, take your candle and your scribe or pointy object. Write your name into the candle—whatever name feels most accurate right now, or you can use a symbol or sigil—and then write or scratch in all your keywords. It's gonna get messy, and that's okay. Just scratch in everything that you have written down.
6. If you want to, you can anoint yourself with oil, and anoint the candle as well. You don't need to coat the whole candle in oil, just a swipe or two will do nicely. Light the candle.
7. If it feels good to you, take the time to call in your benevolent transcestors, protectors, guides, or any other energies that you work with (plants, animals, deities). Once you've called in all your people, read your intentions out loud. After each intention, say some form of sealing words, like "May it be so."
8. Once you've spoken all of your intentions aloud, you can do some divination. This is optional, but a gorgeous time to ask some questions and get some answers. You could ask a simple, "What do I need to know about this process?" Or "What information do I need to be aware of?" Or you can do a spread of cards or similar with more targeted questions. I would suggest:

   card 1: yourself in this moment

   card 2: the energy of the journey you're embarking on

   card 3: one piece of vital information

   card 4: advice from your benevolent transcestors
9. Make sure to record the cards, as well as any insights, into your journal before you move on.
10. Take a moment to shift your attention toward your HRT. Set everything up that you need to take it. Once you're ready, hold the package (vial, tube, pill bottle) in your hand. Take a

moment to reflect on everything that you have done to get yourself to this point. Take a moment to send love, gratitude, and care to all the past versions of you who got you here. Take a moment to open yourself up, mentally, emotionally, and physically, to this next part of your journey.

11. When you're ready, and not a moment before, take your HRT. Whatever form you're taking, savor it. Take your time. Let it be exactly what it needs to be.
12. Finally, when you're ready, take a picture of yourself. You will have this moment to look back on, no matter what comes next. Let the candle burn itself out if you can do so safely. If not, extinguish it and finish burning it when you take your HRT.
13. And hey—congratulations!

Daily HRT Rituals

For some folks, daily HRT is nonnegotiable, either because of how their body functions and responds to HRT, or due to the types of HRT they need for their specific transition. Doing almost anything daily can be a lot to get accustomed to, and for my neurodivergent folks, it can sometimes seem impossible. I don't want these rituals to be exhausting, and I don't want you to feel like you've messed something up if you miss the ritual or find it to be too much.

If you need to revisit chapter 3 and remember the basics for customizing ritual, this is a great time to do that. This ritual, perhaps more than most of the others in this book, will need to be doable for you. Not just doable, but easefully doable: something easy and simple enough that it can seamlessly fit into your daily routine, even when things are stressful or you're in a hurry. So try different things and see what is most effective for your schedule, brain, and energy levels.

Ingredients

- your HRT
- anything needed for your HRT, such as water to take pills
- yourself
- optional: a grounding object of some kind (see the method for details)

Method

1. Prepare your HRT in whatever way you need to. This might be as simple as pulling a tube of cream or packet of gel or a pill out of the cabinet. Just do whatever you need to do to be ready.
2. As you apply your topical HRT or take your oral HRT, take a moment to open yourself up to it, to receive it, to allow it to do all that you wish for it to do in your body. You can also take this moment to offer gratitude or love to your body with the blessing below.
3. Optional: If working with a grounding object, choose an object. Options might include a piece of jewelry, a figurine, a rock or crystal, a framed photo, or an artwork. Each day when you take your HRT, you can touch that object and briefly close your eyes.
4. Whether or not you're working with an object, say this simple blessing, or write your own: "I send gratitude to my body, which shifts and changes. I send gratitude to myself, for my care and love in this transformation. I send gratitude to the transcestors, all who have come before me. Wrap me in your love and your care. Thank you, thank you, thank you."

5. Alternative: If you did the previous ritual, you could say one or more of your intentions, or even their keywords, in this moment.
6. And then move on with your day!

Weekly HRT Rituals

If your HRT is more of a weekly thing, you may have space for a more robust ritual. If you read through this ritual and decide that it is too long or involved for you to do comfortably on a weekly basis, feel free to reference the above ritual and adapt it to your needs. However, if you're craving a more robust experience, read on.

One note: It is common in my experience that folks want a much more robust ritual in the beginning of their journey, and that often their need for ritual wanes as they get farther into their HRT journey. So don't feel that you need to continue performing these rituals weekly for the rest of your life, unless you genuinely want to.

With that said, the fact is that doing your HRT shot can bring up a number of feelings. For some folks, interacting with a needle is scary and brings up a lot of anxiety. For others, doing their weekly shot is euphoric and thrilling. Others may feel their dysphoria spiking at their weekly shot, as it can feel like an unwelcome reminder that hormones are needed in the first place. Whatever your individual reaction to your weekly shot happens to be, ritual can bring some helpful predictability to the equation.

Ingredients

- your HRT
- all the supplies needed: alcohol wipes, syringes, needles, sharps container, adhesive bandage

- your intentions or keywords from the initial ritual (If you didn't do that ritual, no worries)
- journal and something to write with
- optional: a grounding object of some kind
- optional: a candle (Having one big candle for this can be nice; you can light it to begin your ritual and extinguish it at the end.)
- optional: tarot or oracle cards or another form of divination

Method

1. Do at least a brief cleansing, grounding, and protection practice. This can be as simple as a few breaths, wiggling your toes, and drawing a bubble around yourself, or as complex as you prefer. Light your candle.
2. Lay out all your supplies and prepare them in whatever way you need to. Sit with your intentions and see if any of them feel outdated, or if you want to add anything to them. If you want to add anything, please do! If there are intentions you're ready to release, you can draw a line through them, or simply say, "I release you with love." If you didn't do the initial ritual, you can use this time to set an intention—either one for the week ahead, or for yourself and your transition more broadly.
3. When you have settled on the intentions you want to bring in, go ahead and speak them out loud (or in your head, if privacy is limited). If you're working with a grounding object (a rock, a piece of jewelry, a picture) touch it with your fingertips and spend a few moments opening your body up to receive the HRT. If you can, celebrate your body's ability and capacity to change, and yourself for doing what was necessary to get this far. You can celebrate yourself in any way that feels good to you—anything from a gentle acknowledgment to a mini-dance party.

4. If you're working with cards, you can pull a few now. You can keep it simple and ask, "What do I need to know today?" Or you can do a more robust spread of cards. If you're doing a spread, I recommend doing the same spread each week. One example would be:

   card 1: myself in this moment

   card 2: what I should focus on this week

   card 3: advice
5. You can write down the cards you pull and journal any insights that come to you.
6. When you're done with this, take a deep breath or two. Then, do your shot. When you're done, send gratitude to your body, to the transcestors, to the hormones, and to anyone or anything else you want to. If you can, feel the beauty of what it means to create yourself in your own image. And then put all your things away and move on with your day or evening.

Ritual for When You're Feeling Too Much

HRT is a complex and unpredictable experience, and the impact it can have on your emotional life is profound. I find that this impact of HRT is often overstated, yet misunderstood. Clients will come to me terrified that they will become rageful monsters on T, or that they will never stop crying on E. These extremes are unlikely, and yet many folks are surprised by how deeply their emotional life shifts when they start taking HRT. The way I think about this is that your emotional map changes—that is, your emotions shift into new parts of the body, and that paths to them feel unpredictable because they are new.

The path to tears may indeed be shorter on estrogen. The path to rage may be shorter on testosterone. But these emotions are not the only ones that shift, and they are not as predictable as the hype would suggest. Personally, I cry much more on T than I ever did before medical transition, and cultivating a healthy relationship with my anger has been the work of a decade or more.

So this ritual is not meant to be a ritual for any specific emotion or any specific hormonal environment. It is a ritual for processing emotion, for moving it through the body, for surrendering to the experience of it, and for engaging with it in a way that feels safe and helpful for you, specifically. As always, this ritual is one that can and should be customized to your specific circumstances. If the big emotion you're experiencing is white-hot rage, you might need to move differently than you would for grief or frustration or euphoria or whatever else you have going on. Read through the ritual, then customize it to your needs.

Ingredients

- bowl: a large soup bowl, serving bowl, or a mixing bowl is good for this; something large enough that you can put both hands in
- water
- sea salt
- candle: a white or black tea light or chime would be ideal
- paper and something to write with
- towel to dry your hands

- optional: dried herbs. Feel free to choose intuitively or use what you have on hand. Some specific suggestions:
    - for general use: yarrow, rosemary, or dandelion root, leaf, or flower
    - for rage and anger: dandelion root, burdock root, licorice root
    - for comfort or sadness: motherwort, rose, lemon balm
    - for anxiety: skullcap, milky oats, verbena
    - for heartbreak: hawthorn, passionflower, rose

Method

1. This ritual is a little different because we will both start and end with energetic hygiene. Begin with whatever cleansing, grounding, and boundaries feel good to you. The specific purpose of this first round of energetic hygiene is not to cleanse away your feelings but rather to cleanse away any energetic "noise." So as you're cleansing, focus specifically on cleansing away the energy, emotions, and thoughts of others. Ideally, you can come to this ritual very clear on what is *your* emotion and what is not. I recommend you do the breath cleanse from chapter 4, put your feet on the ground or on a solid surface if you can, and put up a bubble to hold your energy and emotions through this practice.
2. Fill your bowl about halfway with water, and then put about a tablespoon of salt in. You don't need to measure or anything, just put a good scoop in. Sprinkle your herbs, if using, and ask them to support you in this work. For example, "Hawthorn, please support, heal, and protect my heart."

3. Sprinkle some herbs on your candle, if using. Place it on the opposite side of the bowl of water from you and light it. Close your eyes if that feels right, or you can gently focus your gaze on the candle's flame. If it feels good for you, take a few breaths into your belly. Try to locate the emotion that feels too big or too much. Touch into it with your awareness. If you can locate any specific sensations in the body, take note of them. Try to keep breathing while you touch in with this emotion.
4. As you feel ready, start moving this emotion toward your hands. You may feel tingling or burning sensations in your body, or some other feeling. This is totally fine; just focus on moving the emotion up the body, and out into your hands. You can start from your toes, and then move slowly up your legs and thighs, your hips and pelvis, your belly and back, your chest and shoulders, then down from your crown through your face and neck. Then push all of that out toward your hands.
5. When you feel ready, plunge your hands into the bowl of water and stir the salt and herbs. You can say the following, or speak your own words from the heart:
6. "[Name of your emotion, such as rage, anger, grief, anxiety, overwhelm], I see you. I know that you are with me. I know that your presence is valid, sacred, and true. [Emotion], express yourself now, through my hands, through my skin, through my pores. Flow into this water and meet the support of my allies, [name the herbs you've used].
7. "[Emotion], through water and salt, be cleansed."
8. You can draw yet more emotion up from your body and through your hands. Like lancing an abscess, draw out as much of the emotion as you can in this way, pushing it through your

hands and into the water, swirling the water with your hands. If you want to make noise during this part, do so! Sob, groan, yell—whatever feels right for you is perfect here.

9. When you feel complete, stop and try to get quiet and still. Focus your eyes on the flame of the candle. Try to find a steady rhythm of breath in your body that feels sustainable and good. Scan your body. If any overwhelming emotions remain, send them to the flame. You can say, "Fire, please transmute this energy into a positive force. Send it where it is needed most."
10. Remove your hands from the water, dry them, and place one or both of them over your heart. Close your eyes if that feels good, or focus on the flame. If there are still flickers of emotion, send breath to them, or imagine the gentlest light enveloping them with compassion and care. When you're ready, take the bowl and pour the water down the toilet. Flush it, wait for the tank to fill, and flush it again. If it feels right, you can say, "I release you."
11. Let the candle finish burning if you can do so safely. Or you can blow it out and redo this ritual with the same candle as many times as you need to.

Ritual for When You're Not Feeling Enough

It is common that folks on testosterone will find their emotional world downshifting into something less intense than what they experienced in an estrogenic body. While this can be a profound relief, many of my clients deeply miss the ability to cry and experience emotional release. This ritual is meant to help create ways to release grief and stress, the same way that tears often do, but for those who are unable to access tears.

Similar to the changing emotional map I described for the previous ritual, I find that testosterone can almost feel like insulation that creates space between our emotions and our awareness of them, or our ability to express them. This does not mean that folks with testosterone-dominant bodies are unemotional, unfeeling, or bad. But hormones do often effect our ability to engage with and express emotions. If you transition any time after puberty, you will be learning to run a completely different programming in your cells. It's not a small thing, and the impact of it hits across every aspect of your life. So be gentle with yourself if you're struggling to get in touch with emotions that used to feel accessible.

Other things can also impact our ability to feel. The dissociation common for trans folks can be at play here, or other types of trauma can also impact our ability to feel and to emote. Many folks are given the message at some point that they can't express their pain, thoughts, or feelings and will often experience this as a block in their throats or chests. So even if HRT is not a part of your journey, this ritual could be useful.

Note that if you're not connected to your body or emotional world due to trauma, any sort of somatic or embodied experience can be incredibly overwhelming. Read through the instructions and then take the experience slowly. If at any point you find yourself overwhelmed, pull back. You can do this ritual as many times as you need to, and you can adjust it as needed.

Ingredients

- bowl, large enough to dip your fingers into
- water

- rose petals (food-grade)
- hawthorn: dried, cut, and sifted herbs or tincture
- oil that works for your skin
    - I specifically recommend an oil with rose in it, but if that isn't available, just focus on getting an oil that works well for your skin. Jojoba is good for this and is readily available.

Method

1. You will begin with energetic hygiene. Cleanse, ground, and boundary yourself as desired, but focus specifically on cleansing away any blocks to your emotional or physical body. Ground into your toes, into your connection to the earth. Rub your hands together and place them on your chest, and ground into that sensation. Put up whatever boundaries you need to feel as safe as you can in your body.
2. When you're ready, fill the bowl about halfway with cool water. Sprinkle the rose petals and hawthorn into the water. Sit with the bowl of water before you. Out loud or in your mind, call out to your emotions. If you're specifically seeking tears or grief, call out to your grief. If you're specifically seeking another emotion, call out to that. Ask the emotion to come forth and make itself known to you, then pay close attention to what happens next. Do you feel some sensation in your body? Do you hear anything? Are there any specific thoughts or images popping up for you?
3. Warm your hands by rubbing them together, then place them either on your heart or on whatever parts of your body were lighting up in the previous step. Feel the warmth passing from your hands into the body, and open yourself up to it. If it feels good to you, take some deep breaths here, receiving the breath and receiving the warmth simultaneously.

4. Take some of the oil on your hands and massage it into whatever part of the body was lighting up, or into your heart or belly. As you massage, gently ask your grief or other emotions to come forward. If you feel like the emotions are hesitating to come forward, ask them specifically what they need from you to feel safe enough to come forward. If it's something you can do, do it! And if not, explain what you can do and go from there. If other emotions or parts of the body come forward, massage oil into them.
5. When you've connected to any body parts that feel like they are holding emotion, and when the oil has been absorbed off your hands, turn toward the water. Use a hand to stir the rose petals or hawthorn into the water, focusing on stirring the energy of compassion and acceptance into the water as well. When you're ready, use your fingers to drip the water down your cheeks. These are your tears, and you can repeat this as many times as you want to. You may find yourself actually crying in this process, and that's fine. It is also fine if that doesn't happen for you. You can elect to hum or make noises, or if that doesn't feel good, just take long, slow breaths deep into your belly.
6. When you feel complete with the water, close your eyes or let your gaze go soft and breathe in gratitude to your body, your heart, and your emotions. Send gratitude to yourself and to the water and herbal allies that were present for this ritual. Take some time to journal about your experiences if that feels good to you, then move forward with your day.

11

Rituals for Surgery

Of all the various steps involved in transition, surgery is a natural fit for ritual. You wash and prepare your body, you submit to death (anesthesia), and you emerge reborn. But the ordeal is not over when you wake up in a hospital recovery room. You still have the process of recovery—watching your bruises darken and fade, tending carefully to the needs of your body. For some surgeries, recovery is a brief process. For others, it can take many months. In all ways, each surgery is a threshold moment in life, and each surgery offers a perfect opportunity for engaging in ritual.

Folks have many different relationships to their pre- and post-surgery bodies. When I went for top surgery, I was absolutely thrilled to be having my chest tissue removed. I had no complicated feelings about it at all, aside from "What if it doesn't happen?" However, as a gender doula, I've had clients who felt any way you can imagine before gender-affirming surgery. Clients who needed to process grief associated with the part of their body where the surgery would take place. Clients who needed to process what it means to alter their body. Clients who had lived many decades in their body and felt complicated feelings about what it would mean to willingly change it.

I've worked with clients who were accessing a revision for a gender-affirming surgery that went awry, and clients who were getting surgeries to effectively "reverse" a gender-affirming surgery that no longer felt right to them. I've worked with clients who were absolutely terrified of their surgeries but were brave enough to pursue them anyway, and clients who had been doggedly working toward accessing their surgeries for years despite many struggles. I've worked with clients who felt very isolated and alone in their surgical journey, and clients who were surrounded by family and friends and support.

In all of these situations and in each layer of complexity, ritual was a helpful tool. This chapter will provide some of the most common surgical rituals I'm asked for, but your circumstances may not perfectly align with what is written here. As always, take what works, remix it, mold it, make it your own, and leave the rest. While I've organized these into separate rituals, the fact is that in most surgical journeys, all of the rituals in this chapter will likely be relevant. Read through the included rituals and see what works best for you.

Ritual to Prepare the Body

In a surgical journey, your body will experience some level of shifting and changing. Sometimes, parts of your body will be removed. Sometimes, they will be refashioned into a new shape, function, or purpose. Sometimes, new pieces will be added to change the function, appearance, or meaning of the flesh you've got. Regardless of what exactly will be happening with your body, it's profoundly helpful to create a ritual to support your body through the shift, releasing the parts of you that need releasing with love and care, calling in the new shapes, parts, or functions, and most of all, opening a dialogue with your body about the fact that surgery is happening, or is desired, and making space for that to be a conversation.

If you read or conducted the HRT rituals from chapter 10, this last bit is similar to the ritual to prepare the body for HRT. We want to move with our body in a consensual way, when and where possible. Again, I believe that the distress or dysphoria that we often associate with our bodies is our body making an effort to communicate with us. While the idea of going under anesthesia and having your body cut open can be distressing and bring a lot of other body-based feelings (anxiety is often a big one), the fact is that if there is either a desire to change your embodiment, or distress at your current embodiment, or both, the call is coming from inside the house. Your body wants to change, it is asking to change, and you can make it your most powerful ally in this process of becoming.

So we start here, from a ritual to prepare the body. This ritual will be largely composed of grounding in and starting a dialogue with your body, so the main tools that you need are those that help you do this. Your breath is a great place to start. If you use any system of divination, this is a great place to incorporate that. I will suggest some prompts you can answer with cards, which you can adapt any way that works for you, but you can also use them as journal or contemplation prompts. Try to do this ritual when you have ample time for connection and reflection. If needed, you can do a shorter version of this ritual daily or weekly in the time leading up to your surgery date. Read over the method, and adapt it as necessary.

If you have not yet done the rituals from chapter 6, they will help with connecting to your body in this particular way, so I recommend doing them first.

Ingredients

- optional: journal and something to write with
- optional: oracle or tarot cards, or another form of divination

Method

1. Do whatever energetic hygiene you need to in order to prepare yourself and your space. This ritual is best done while sitting, so find a comfortable spot, cleanse it, set up your boundaries, and settle in. It is best not to be distracted during the ritual, so set your devices on Do Not Disturb, or better yet, put them in another room, and take whatever steps you need to be present and tapped in.
2. You will begin with connecting slowly with your body. You can do this in any way that makes sense to you. Some options include:
    - Take five deep, slow breaths into your belly. Breathe in through the nose if you can and out through the mouth. Pause and see if you want to take more. If you're ready, move into the next part of the ritual.
    - Take time to settle into the space where you are, feeling the weight of your body on the surface where you are sitting or lying. Let yourself get as heavy as possible, then begin to relax your body one part at a time, from the top of your head to the bottom of your feet.
    - After settling into your space, tap lightly on the top of your head. Let your tapping fingers move down the sides of your face or across your face, all over your head and down your neck, to your shoulders and collarbones and chest. Tap on your belly and hips and down your thighs and calves. If you can, tap your feet or wiggle your toes.

    Or choose another method that works for you.
3. Once you're settled into your body, take time to put particular focus into the parts of you that will be changing. You can send

breath to those parts, or visualize a specific color or quality of light flowing into them. You can send them energy, care, or anything else that feels good to you. Essentially, you are saying hello and opening a space for dialogue with them.

4. Start by asking these parts of you, "Is there anything that you need me to know?"
5. You can wait for a thought, feeling, or other form of communication, or you can take this moment to pull a card and see what it says. Whatever information your body imparts to you, make sure to send gratitude, care, and love back to your body to say thank you. If you want to, you can write down the card and your interpretation, or anything else that you want in your journal.
6. When you're ready, start to inform your body of the changes that you're working on. Only you can know the best approach for your specific self, so move slowly and do what feels best to you. One option is to start by recalling any desire for changing or shifting specific parts of your body, or recalling any distress or dysphoria that you feel about the body that you have. You can start from that point, letting your body know that you're addressing that desire or distress directly.
7. Now, see if you can allow yourself to receive whatever response your body has for you. If your body gives you distress signals, this is okay. Try to be with it. Feeling anxious, upset, grieved, or any other hard feeling isn't a negative thing in itself. Letting that feeling be with you, and trying to open up to it more and more, will help you to process whatever it is that you need to process before surgery. You might also feel joyful, relieved, at peace, or nothing at all. Whatever happens, focus on allowing it, accepting it, and letting it be.

8. Once you've allowed yourself to feel anything that comes up, you can pull another card with the prompt, "How am I (or how is my body) feeling about this right now?" If you want, take this time to journal your reflections.
9. Finally, you can settle back into your body and ask, "Is there anything that you need from me to help prepare us for this process?"
10. Start by being open to any feedback or information that your body has for you. This might be in the form of thoughts, feelings, sensations, ideas, snippets of a song in your head—there's no wrong way for your body to communicate with you, so just be open to whatever happens. Again, let yourself feel whatever there is to feel at this point. Then, if you would like, you can pull a final card. I recommend asking, "Is there anything else I need to know?" Take the time to journal or reflect on whatever arises.
11. Send gratitude to your body. If you can, gently touch your heart, and whatever part of your body is going to have surgery, and send extra love to those places. Check back in regularly if you can, and see what shifts and changes over time.

Ritual for Medical Anxiety

Medical anxiety is very common among my clients, and there are a number of ways to work with it. First of all, to be very clear, you're always able to talk to your surgeon about medical anxiety. Most surgeons will prescribe antianxiety medication for the morning of surgery, though you usually need to fill out your consent forms before you actually take the medication so that they can be sure that you know what you're consenting to. If you find you're dealing with a lot of anxiety before surgery, you can talk to your primary doctor or psychiatrist about additional medications for that specific time. If antianxiety medication is not for you, there are many herbal allies that help to alleviate anxiety. If you turn to herbal support, make sure your surgeon knows what herbs you're taking. Some of them need to be stopped two weeks prior to surgery.

But in addition to this, ritual can be incredibly helpful with presurgical anxiety. My most common recommendation to clients is to use a ritual to create what I call an anchor phrase. This is a phrase or a series of phrases that can be repeated again and again, either out loud or in your head. Repeating phrases again and again can have a soothing quality. It can help to regulate the breath and bring a sense of calm to the body. Some folks might use prayer for this—a common prayer I rely on is the Orphic Hymn to Asclepius, which has seen me through many a medical crisis. But prayer is not comfortable or available for everyone, and an anchor phrase has the benefit of being customizable to exactly whatever it is that helps to anchor you.

Some folks might use the word *mantra* to refer to this type of practice, but mantras serve a specific and sacred purpose in Hinduism, so I don't use that term broadly for this purpose.

Creating an anchor phrase can be done as an addition to the previous ritual, especially when you've done it more than once. Consider doing the ritual to prepare the body once a week for a month or moon cycle, then on the final instance, add this ritual at the end. I've outlined a few different methods for generating this phrase in the following ritual, but if none of these resonate with you, that's not a problem. Do whatever makes the most sense to you and works the best for your needs.

Ingredients

All of the ingredients are optional. Read the ritual first and decide what you might need.

- journal and something to write with
- lotion or body oil
- music and some way to listen to it
- books of wisdom, prayer, or poetry
- magnetic poetry

Method

1. Start with cleansing, grounding, and setting up your boundaries, both on your person and in the space that you will be in. Focus on clearing out any energy that isn't yours, that isn't helpful, or that doesn't serve your purposes. Take time to call in any helpful energies, guides, or supports. I recommend calling in the transcestors, especially those who have experienced surgical transition and who want to help and support folks who are seeking that form of transition for themselves. Also feel free to call in the support of any plants, animals, or other energies that you work with regularly or find supportive.
2. When you've opened the space, take the time to get in touch with your body. This doesn't have to look any particular way. I recommend putting on some music that you enjoy and moving intuitively to it. You could also break out your favorite lotion or body oil and massage it into all the parts of you that you can reach, or that you feel like engaging with. The goal of this part of the ritual is to connect with yourself, with your body, and to use that connection to receive some embodied wisdom.

3. When you've gotten nicely connected to your body, you can ask, either out loud or in your mind, for a phrase to support you as you move through surgery. You can request this directly from the transcestors, or from the various powers that you called in at the beginning of the ritual.
4. At this point, you will need to do what works for you. Here are some options that I find helpful.
    - Grab a book of wisdom or some helpful text. Flip through it until you feel like stopping on one specific page, then drag your finger around the page until you feel like stopping. Do this until you have a set of words, then move them around, add necessary modifiers, and create a phrase that works for you.
    - Get quiet, close your eyes, and focus on being present in your body. Wait for the words or phrases to float up into your consciousness and write them down. If they come as disjointed words, experiment with order until you come up with something supportive and helpful.
    - Read through poetry or prayers (Sara L. Mastros's *Orphic Hymns Grimoire* is great for this) and either choose one that you like, or use them as inspiration to generate a phrase or prayer of your own.
    - Use magnetic poetry to put together a few words, and fill in the rest intuitively.
    - Think about words that help you feel strong, grounded, and calm. Use these to create the phrases that work for you.
5. Ideally, you will come to land on a few words, a phrase, or a few phrases that feel supportive and helpful. Practice saying these words during moments of calm and relaxation at first.

If you have time before surgery, work with the words during times of stress and anxiety as well. Find a rhythm to say them that feels soothing to your body. Memorize them if at all possible. Try saying them quickly. Try saying them slowly. Try letting them drip over your tongue and roll around your mouth until they lose all meaning, then keep saying them until the meaning returns. Anchor phrases become more and more powerful when we repeat them, because we imbue them with our energy and find different ways to feel them and engage with them. So work with your phrase, make it your own, and when you need it, it will be ready for you.

6. When you've found the words or phrase you will be using, send gratitude and thanks to all who you called in, and close out the ritual.

Ritual for the Possibility of Postsurgical Regret

One of the many ways that anxiety shows up around surgery is the fear of postsurgical regret. As I am writing this in 2025, medical transition regret remains a favorite talking point of transphobes everywhere. Because they vastly overblow the idea of it, folks who are seeking surgery find themselves in a strange situation. Being honest and authentic about their fear of regret feels impossible, since it could be mistaken as internalized transphobia, but ignoring those fears and pretending they don't exist isn't particularly helpful.

So let's be very clear. The likelihood that you will regret surgery is very low, but it is not zero. There are people who regret their gender-affirming surgeries—I've doulaed some of them! The fact that regret happens should not be some boogie monster that we don't talk about. Regret is a fact of life. People regret all sorts of things, and it's not always clear when we embark on a new path whether or not we will come to regret it. In fact, you will see folks strongly encourage folks to do things that other people regret all the time. Getting married and going to grad school are my favorite examples of this. The number of people who regret either of those choices is high, yet folks still encourage these paths without thinking twice.

I say all of this to take the mystique out of regret. Regret is simply a thing that happens sometimes as a consequence of living. The more you do and experience, the more likely you are to land on something that you regret. And also, each experience that you have, each action that you take, creates an opportunity for you to learn and grow. To me, regret is a side effect of growth. And even the things I've regretted in life have been potent teachers. The clients I worked with who have regretted certain parts of their gender affirmation journey have also come to appreciate the experiences that they've had. Each of them has found meaning in that experience or has found agency and strength in their ability to move forward from it. Each of them is stronger and more self-assured than when they came to me, not because of me, but because they took the time to cultivate the gifts of their experiences, good and bad.

And also, if you're presurgical and feeling afraid of regret, all of the information above probably won't soothe you for long. Even if you know these things in your mind, it's hard to get them into your body. So this ritual is meant to help you do so. I don't find it particularly helpful to try to shove hard feelings away or pretend they aren't there. Even dissociating from the hard feelings may only be effective in the short term. When we push our feelings down, back, or away, they often come back stronger and more powerful than ever. So in this ritual, we instead make space for the feelings.

Ultimately, your fear of regret is some part of you that wants to protect you from every bad outcome. And that is valid. But regret is not the worst outcome, and in order to do the things we need to do in life, we simply must take risks.

So, take a deep breath, and remember that you have gotten through challenging experiences before, and you will again. In the meantime, this ritual will support you.

Ingredients

- candle of any kind; a tea light or chime is recommended
- paper (several sheets, loose or easy to tear out of a notebook), and something to write with
- ribbon or string

Method

1. Take time to cleanse away all that doesn't serve you, that isn't yours. Focus specifically on cleansing away other people's stories or opinions, unhelpful assumptions, and the anxieties that aren't yours. It is okay for your emotions, stories, and anxieties to be present, but you don't need to take on anyone else's, especially while you're doing this tender work.
2. For grounding, I invite you to specifically ground into the self who knew they wanted the surgery. There is a past version of you who made the effort to schedule, to figure out insurance or how to pay out of pocket, who did the necessary research and effort to make this surgery happen. Even if you felt unsure at the time, some part of you was sure enough to schedule the surgery. Tap into that self, and ground into their energy.

3. For boundaries, draw your bubble such that you call in your benevolent transcestors, supportive energies and archetypes, anything protective and helpful. I specifically recommend reaching out to the transcestors who have had surgical transition and who know what you're feeling in this moment. They have wisdom and they are generally quite happy to share it.
4. When your energetic hygiene is complete, light a candle and set it in front of you, far enough to make room for the paper and for you to write. Focus your gaze on the flame, then allow your gaze to soften, and for the flame to blur and dance in and out of focus. While letting your eyes remain on the flame, put a hand on your belly and a hand on your heart. Call forth the stories, anxieties, and fears that are in your body. If you are feeling any sensations in your body, move your hands to that area. Don't try to change, push away, or make the hard feelings better. Just offer your own touch as support and care.
5. If you're having trouble getting in touch with your fears and anxieties, bring up the idea of postsurgical regret and see what happens. You don't want to be completely overwhelmed or flooded—if you're feeling that way, ask the feelings to titrate themselves. You can say, "This is too much, can you give me just 10 percent at a time?" Or something similar. You can also just ask them to step back a bit and give you a little more space.
6. As you hold these fears and anxieties in your body, do your best to take five deep breaths. With each breath, focus on making more space for your fear, even 1 percent more space. I like to think of the breath as opening up space in my ribs, and that space making room for the fear and the anxiety or the hard feelings to live in my body. Let your eyes focus on the flame; let yourself breathe open more and more space.

7. When you're ready, take your paper and something to write with. Write down, to the best of your ability, all of the fears and anxieties and worst case scenarios that are bothering you. As you write them, even if they feel or sound absurd, try to focus on holding them with a gentle sense of compassion. Remember that these parts of you are doing their best to protect you. Remember that they are scared and that it is okay to be scared. Try to orient to them as you would a young scared child or pet. Calm and gentle above all.
8. When you've written down the fears and anxieties, read them into the flame one at a time. Literally speak them aloud, even at a whisper, with the candle flame as your focus. You do not need to physically burn the paper, just allow fire to take the fears, anxieties, and stories, and to cleanse them as you speak them aloud. When you've said them all, send gratitude to the fire, and send gratitude to your fears and anxieties and the parts of you that want to protect you. Hold your hands a safe distance above the flame, close enough to feel the warmth but not so close as to hurt, and let it warm them. Then rub your hands together to create friction, and press them against the part of the body where your anxiety shows up. Do this as many times as you need to until you've brought warmth and support to each anxious part.
9. Now take a fresh sheet of paper and write out the best possible outcomes. Do your absolute best not to restrict them, even if they feel unlikely or unreal. Try to write them in the present tense, as if you've already completed your surgery and you're thrilled by it in every way. Phrases like, "I adore my gorgeous chest" or "My new vagina feels incredible" or "I love my face

more than I ever dreamed I would." When you've written as much as you can, take the time to read these out loud to the candle. Instead of sending these to the flame to be purified, send them out as intentions. These words are a spell.

10. Roll the two pieces of paper together and tie them with a string or a ribbon. Put them away, either on an altar you've made for your surgery or in a place where they won't be disturbed.
11. Finally, write a letter to your future self. Write down all the reasons why you're pursuing this surgery. Write down your hopes and plans. Write anything that you want your postsurgical self to know. As much as you can, send them love, support, and care. Put this letter in an envelope and address it to yourself. Across the flap, write, "Open when you need to." Place it on an altar, or someplace where you won't forget it.
12. To close this ritual, thank your candle, yourself, and all those who you called in. Extinguish the candle.
13. As you move through the portal of surgery, remember the letter. If you need support, care, or a reminder of why you did this, open it and read it. You can read it as many times as you need to. Once you've recovered from surgery, open the roll of papers and read the two lists. Reflect on the you who wrote them. If there is grief to be felt, try to use the breath to make room for it. If there is joy to be felt, you can use the breath to make space for that as well. Most of all, reflect on the person who wrote those lists. Reflect on all the versions of you who wanted this surgery, sought this surgery, passed through this surgical portal, and made it to the other side.
14. Celebrate yourself, as much as you can, in every way that you can.

Ritual to Bless Your Surgery

If you've made it to the week of surgery, you're ready for this ritual blessing. Congratulations!

You do not need to be in any particular state of mind for this ritual. By this, I mean that if you're feeling unsettled, anxious, excited, scared, or any other challenging emotion, you can still do this ritual. I will invite you to cleanse, ground, and boundary yourself as always, but that does not mean that you can't do the ritual. Chances are, if you're this close to surgery, you're feeling all kinds of emotions. All of them are fine, and none of them are a problem. They can all ride along with you while you call in all of these blessings.

Although each of the rituals in this book is modular, and although you can customize any and all of them to your heart's content, this ritual is particularly available to be customized. You can bring in an officiant to conduct this ritual or do it on your own. You could do this ritual with friends and beloveds too. If you want to call in the blessings of other energies, deities, specific transcestors or ancestors, please do! When I originally wrote the blessing for this ritual, I wrote segments of it specifically to the energies and deities I personally work with. When I say the blessing, it is with those parts included. When I give the blessing to others, I leave those segments out and encourage them to add their own. I'll encourage you to do the same.

If you want to, you can record yourself reading the blessing in this ritual, and play the recording back to yourself on the day of surgery. This can help ease anxiety, and also, there's no problem with calling in a maximum amount of blessings.

This ritual uses candle magic and the blessing itself. It's not overly complex, but you can add anything that you want to it if you need it to be more robust or just different in some way.

Ingredients

- candle of any type
  - This might be a good candidate for a larger candle, especially if you have a way to burn them safely instead of extinguishing them. If you don't, just stick with a tea light or chime.
- something to scratch the candle wax with: a scribe, tack, or ballpoint pen that you don't mind getting waxy
- blessing oil
  - This can be any oil that you like, an oil of cultural or spiritual significance, or an oil that you or a friend made specifically for this purpose.
- dried, cut, and sifted (or powdered) herbs for healing
  - recommendations: rosemary, rose of Jericho, calendula, nettles, oat straw, dandelion, rosehips
  - None of these are to be taken internally for the purposes of this ritual; you just need a very small amount to dress the candle.
- list of words that represent the best outcomes from your surgery, or sigils

Method

1. As always, begin with the energetic hygiene that you need for yourself, your space, and anyone else joining in this ritual (see chapter 4). Tip: If this is being done with more than one person, do the energetic hygiene together—cleansing, grounding, and boundaries—as it helps to sync up your energies and align them toward the goal of the ritual.

2. If folks on your care team, or anyone else who is prominent in your personal sphere, are experiencing a lot of anxiety or worry on your behalf, it can be helpful to take extra steps to cleanse their energy out of your personal bubble. If they are amenable, they can work on making their own bubble or energetic boundaries to hold the anxious energy. If they are not, you can focus on putting their energy in a bubble separate from your own.
3. Once you have prepared your energy, settle in and dress your candle. You want to add your name to it, as well as the time and date of your birth if you have it. If using a tall type of candle, scratch these details in starting at the bottom and moving toward the top. If using a pillar or tea light, you can scratch these details around the edges in a circle. Then, add words that represent the best possible outcomes. You can write words in the language of your choice, but you can also use these words to make sigils if you'd like (see chapter 5).
4. When you've scratched the words or sigils into the candle, brush the bits of wax away gently and then use your finger to anoint it with oil. Again, for a tall candle, draw your finger with the oil from bottom to top. If you're using a tea light or pillar, draw the oil around the entire top of the candle. Then either roll the candle in herbs or sprinkle a light dusting of herbs onto the top of the candle. (Note that candles with herbs in them burn a little differently and can be unpredictable, so monitor the candle while it is burning.)
5. If you have not already, take a moment to build an altar for the surgery, or at least for this working. If the altar needs to be portable, you can create it in a box or tin. At the very least, put your candle and a cup of water on it. If you can add additional offerings, a picture of yourself, something to represent your surgery or your transness, do that as well.

6. When you're ready, light the candle, and take a moment to focus on the flame. You can keep a sharp focus on it, or you can let your gaze get soft. While you're gazing at the candle, call in transcestral support or any other helpful and benevolent support and aid. Let yourself settle into your body, and touch in with the part of you that is getting ready to shift. And then, when you feel ready, either say the following blessing or have it read over you.

Gender-Affirming Surgery Blessing for the Precipice of Embodied Transformation

To the benevolent transcestors
Imaginal Ones,
You who have gone before,
You who guide and protect us,
You who see the larger story
From your place in the Beyond.

To those benevolent transcestors
Whose bodies have been transformed
To those benevolent transcestors
Who support health and wellness
Happiness and thriving
Who see the brightest future
And guide us toward it.

To all of my guides,
To all of the benevolent spirits who watch over me,
To all of my protectors,
Who come in myriad ways
To save, to soften, to shield, to cushion
To keep me from severe harm.

[If you want to invoke others here, you can.]

To each of these powers, I address my petition:

From each of you
In accordance with your divine gifts,
And in the manner that pleases you,
I ask the following:

That I will have a successful procedure
That all personnel who touch me
Either directly or indirectly
Will touch with the hands of healing
That all who work with me
Directly or indirectly
Will operate at their highest and best capacity.

That there will be no errors
or omissions in my care
That all procedures will be completed perfectly,
That all results will be beautiful to my eye,
That all medication will function only as desired.
That all negative side effects will be minimized
That my body will heal perfectly and well
That I will be surrounded on every side with peace,
That I will be utterly safe
That I will be blessed in every conceivable way
throughout this process.

This is my petition.
I ask for these blessings with love
With care
With deepest gratitude

For these boons I make offerings
of fire, of water, of my breath.
For these boons I make offerings
of my faith and my devotion.

So I have asked, so may it be. [three times]

Take some time to focus on the candle, to feel into your body, to let yourself be present with the energy of the blessing, the surgery, and anyone who has gathered to support you in this. When you feel complete, place the candle somewhere it can safely burn, or extinguish it. Give gratitude to your guides, transcestors, and anyone else that you've called in.

Happy surgery!

12

Transcestor Rituals

One of the most powerful and impactful aspects of my spiritual practice is working with the transcestors. When I say "transcestors," a portmanteau of *trans* and *ancestors*, I am referring not only to trans folks from my bloodline but also something far broader. To get there, let's start from a more general approach to ancestor work.

Ancestor work can be either a spiritual or cultural practice, or both. For those of us (like me) whose ancestors chose to profit from colonization and assimilate into whiteness, ancestor work can be a complex experience, certainly not always lighthearted or joyful. It is powerful work, sometimes quite heavy, but in my personal experience, deeply fulfilling. Many folks who lack experience or training in ancestor work assume that they must connect with all their ancestors, no matter how violent, and that "ancestors" specifically refers to the people whose names and faces they know in their familial bloodline.

This is not my understanding of ancestor work at all. The ancestors that I personally connect to most deeply are folks who lived centuries ago. They are not people whose names or faces I know from records or photographs—they lived before either of those. The ancestors I connect with most readily, and who are most willing to connect with me, are not even always human beings. Before there were humans, there were many other forms of life. And each life form is connected in a web to all other forms of life, and I would add, to all things. This is what I refer to when I refer to the web of all things, and it is a vital and important part of my approach to the earth, myself, my community, and my spirituality.

When approaching transcestral work, my underlying belief system is quite the same. I speak to "the transcestors" as a collective, as a shorthand for all the trans folks who ever were, and all the other things and forms and lives and energies that embody transness in the web of all things. To be clear, a person who lived outside binary gender three thousand years ago certainly wouldn't use the word *trans*, as this is a modern description that only emerged in the last century or so. Yet, there have always been people who lived in the liminal, who transcended binary understanding, who were more than the body they were born into, who shifted their sex. The things that make trans people what we are have always existed. This thing we currently call "transness" is present all across nature, is present in the majority of cultures and religions, and has been with humans as long as we have been humans.

On an episode of the podcast *How to Survive the End of the World*, I heard Geryll (Dr. G. Love) Robinson talking about ancestor work. She said that if all of your ancestors are represented by a huge mighty tree, then the human part of your ancestry would be just one branch of that tree. Our ancestors are far more than the humans whose DNA we share, and our transcestors are as well. Some nonhuman transcestors I work with regularly include Dandelion, Hummingbird, Spider, and Oak (especially Coast Live Oak), among many others. What is the criteria for knowing that a nonhuman ancestor is a transcestor? Ask them.

This type of knowledge is often referred to as personal gnosis—that is, your own deep spiritual knowing that is independent of others, and that you do not try to preach as gospel to others. In the tarot, personal gnosis is represented by the High Priestess—a card about the wisdom that comes deep from your guts and that is specific to you. Not wisdom to be broadcast or shared about per se, but something more personal than that, more private. In my opinion, all deities are trans, and basically any spirit that exists can be trans too. You may not agree with me, and you don't need to. No matter how you view transcestors—human blood relations who are trans, human non-blood relations who are trans, nonhumans who have trans energy, energies that contain transness, or something else entirely—the fact is that you have them. We all do. And we can all reach out to our transcestors for help, support, and guidance.

This topic is way too much to fit into one book. My dear friend Zel Amanzi collaborated with me to create a whole class about this topic. It started as a workshop, but after we presented the material, many of those who filled out our feedback form shared the same complaint: "I wish it were longer!" We took the note and transformed it into a five-week class with experiential learning, optional small groups, a whole meeting just for processing, and much more, yet still our students wanted something longer, more sustained. I say all of this to point out that this is a topic that is well beyond the reach of this book alone.

Yet you don't need my permission to connect with your transcestors. You don't need a class or a workshop or a book. You simply need to start connecting. This chapter is meant to help you start that process right from where you are. If you don't know what you think or believe about transcestors yet, that's actually perfect! You have the opportunity to start experimenting and see where it takes you. Even if you're not sure that transcestors are "real," give these rituals a shot and then pay attention to what happens next. There are many ways that transcestors will show up in your life, many ways they can assist you, communicate with you, and be present for you. Your relationship to them will be your own, because your energy is your own, and your connection to the web of all things is your own.

Even Zel, who vibes with me enough to teach a whole class with me, has a different approach to transcestors than I do—an approach that is both unique to his experience and informed by their African diasporic heritage. There may be cultural or spiritual practices that you already observe where transcestors can be invited in. You may already know the name or face of a transcestor in your actual bloodline, or you might not. Start with where you're at, try the rituals, keep notes, and see what happens.

For me, working with transcestors was what allowed me to access my spirituality. My relationship with the transcestors helped me realize that I was being called to gender doula work, and once I started doing that work, the transcestors also helped me to make crucial choices that I would not have otherwise moved toward. The transcestors have provided me with important words for my community and others that are just for me. They brought me together with Zel. They told me to write this book, and they have informed (or in some cases, dictated) every single ritual in it. I believe strongly that each and every trans person benefits from transcestor work.

The following rituals are meant to help you begin developing your own relationship to transcestors and learning how to ask for guidance and support. This is just a bare-basics primer. There's so much richness to be had, but my hope is that folks will get a taste from these rituals and begin developing their own practice from there.

Honoring, Uplifting, and Venerating Transcestors

Developing a relationship to any being can feel daunting or intimidating at first. There are some pretty standard practices that can help, and while they aren't necessarily complicated, they are quite important. The key here is something that I like to call devotional practice. Devotional practice means showing up, again and again. It means doing the things that you say that you will do—making a commitment to yourself and to something that you care about and then showing up to live out that commitment in real time.

One common form of devotional practice is a garden. If you're growing a garden, and you want it to be prolific, you will need to show up regularly. Maybe not every single day like clockwork, but regularly enough that you can tend to what is growing, pruning what needs to be pruned, watering what needs to be watered, harvesting what is ready to be harvested. This can be very similar to the garden of our interpersonal, spiritual, and energetic relationships. I don't text every single person that I love every single day, but I show up as best I can each day, and if I realize it has been a minute since I checked in, I will do what I can to rectify that. If I know someone is in a particularly challenging time, I will put extra effort into showing up for them as best I know how.

Your relationship with the transcestors (or any spiritual relationships, really) can be much the same. You start by getting to know them. You can offer gifts, refreshments, or anything else that feels appropriate. You can ask them to connect with you, to give you messages, to let you know how they wish to communicate with you. Offer them gratitude, care, and respect, and you will cultivate something that is both specific to you and deeply supportive.

If you've never cultivated a spiritual relationship before, or if your only experience is a source of trauma or shame, this might seem incredibly daunting. This ritual is meant to create a gentle pathway to this sort of cultivation, to reduce complexity and render it simple and approachable. Read through the following ritual, and let it simmer. You don't have to do it right away if you're not ready. Read over it, let yourself feel into it, and then move accordingly.

Ingredients

- items to build an altar (see chapter 5)
    - suggestions: pictures of transcestors (your own blood or spiritual ancestors, or transcestors who have been elevated such as Marsha P. Johnson, Sylvia Rivera, or Lou Sullivan), transcestral herbs and other correspondences (you can choose these intuitively, or use dandelion), items representing the

elements, items representing your own gender or transition, photos or items that represent you, beverages like tea, coffee, or alcohol, and anything else that feels right

- small cup of water
- candle: a tea light works great; any candle will do
- journal and something to write with, or some way to record things

Method

1. Take the time to do the energetic cleansing, grounding, and boundaries that you need to do. The most important aspect of this ritual is focus and care, so cleanse away anything that is distracting you or impeding your focus. Take time grounding into your own energy and into the earth or the elements, and set your bubble such that only those who are invested in your highest and best outcomes can engage in the ritual with you.
2. Take the time to call in your transcestors, specifically your benevolent transcestors. It is important to specify this, especially if you're new to ancestor work. You can say something like, "I call in my well and benevolent transcestors—only those who wish to see my highest and best outcomes, and no one else." If you're experiencing doubt, trepidation, or fear, it's okay to name those things. Let your transcestors know how you feel and what's going on for you, and ask them to meet you where you're at, in a way that you can understand and respond to.
3. Build your altar. This altar should be more permanent if possible, so make it in a space where it can live undisturbed (see chapter 5 for more details). You want to be able to come back to this altar regularly for this ritual as well as the others in this chapter. Name each item as you add it to the altar, and name why you're adding it. For example, "I place this feather on the

altar to represent the element of air" or "I am adding this photo of myself because this altar is a connection between me and the transcestors." Give as much or as little of a description as you feel comfortable. Even something like, "I am adding this wooden button because it felt like an important thing to do," is fine.

4. Finish by adding your cup of water and light the candle. Pause and take some time to settle your body, either with breath, intuitive movement, or whatever works best for you. When you're ready, settle your gaze on the altar. You can use the candle as your focal point.
5. Speak this next part out loud if you can. If you must, say it in your head.

   Beloved benevolent transcestors,
   You who have lived in the liminal
   You who have crossed between
   And among categories

   Beloved benevolent transcestors,
   You who watch over me
   Who look after me,
   Who wish to help me find
   My highest and best path

   Beloved benevolent transcestors,
   I call to you.
   I made offerings to you
   Of flame and water
   Of [name anything on the altar here].

   Beloved benevolent transcestors,
   I ask that you join with me
   Clear the way and open my path
   Protect and guide me

Help me find the highest expression
Of my purpose.

Beloved benevolent transcestors,
Hear me call to you.
For I am like you,
And you are like me.

Beloved benevolent transcestors,
My gratitude to you.
My care to you.
My love to you.

6. When you've said these words, take time to be still and see or feel what happens next. Do you sense any specific energy? Any sensations in your body or in the space around you? Do you hear anything or feel any ideas coming into your awareness? If you're new to this type of intuitive sensing, you may not be sure if something is transcestors or just random brain noise. Don't worry about it too much. Notice anything that is present and make note of it in your journal or in whatever way works for you.
7. When you feel complete, make sure to send gratitude to yourself for showing up, as well as to the transcestors. If any specific transcestors came through to you, make sure to thank them and offer gratitude.
8. **Now, perhaps the most important part.** Continue to come back to this altar regularly. Make offerings here daily, or at least as regularly as you can. Offerings can consist of a small cup of water, plants or flowers that you find in the day, small portions of your meals or snacks, tea or coffee, alcoholic beverages, incense, or anything else that you want to offer. If all you have

> time for is to place the offering, do that. But if you can, take the time, even if just a minute or two, to get quiet and connect with the transcestors. You can say the prayer again if you like, or a shorter version. You can also just say "Hello!" when you think of it.

Remember, this is a relationship, and it is normal that it takes time to build. Pay close attention in the days, weeks, and months following this ritual to see if you start noticing shifts in your dreams, your thought patterns, or specific feelings in your body. Pay attention and see if you are consistently noticing a certain animal, phrase, number, and so on. The transcestors communicate with me in a variety of ways, and when I talk to others who have these practices, I am always amazed by the different ways these relationships show up in our lives. When I work with clients around transcestral practice, they often ask, "How do I know it is the transcestors and not something else?" This is a great question. First of all, you're specifically inviting the transcestors with the invocation written above. Similarly, you can focus on only inviting or accepting benevolent transcestors and no others. Use your energetic hygiene work and the skills you've been building throughout this book to clarify and discern the energies you're experiencing.

Ultimately, though, there is a large part of this experience that is just trust. One might say "faith," although that word can be tricky for me personally due to my religious trauma. I've found my own ways to trust my knowing, to trust my read of different energies and influences, and a lot of that has just come from consistent effort. Guessing and checking. Trying different things over time and experimenting. You will find your own way into what works for you, specifically, and it doesn't need to look like what works for me. But if in doubt, try, and see what happens next.

Connection with Specific Transcestors

The previous ritual is about connecting with the benevolent transcestors more broadly, and I recommend completing that ritual first. However, some of us may have known transcestors in our blood line, or may be interested in connecting with a specific transcestor like Marsha P. Johnson or Lou Sullivan. Zel taught me that folks like this can be referred to as "elevated ancestors," that is, ancestors that are well-known and that can be honored and worked with by many folks. You may also be interested in working with transcestors who share your specific experiences, such as transcestors who have accessed gender-affirming care, or transcestors who weren't able to access this care.

There are ways to work with each of these types of more specific transcestors, and for each, I recommend starting from a working transcestral altar. We made such an altar in the previous ritual, and I recommended that you create an altar that you can both leave up and return to when needed. Altar work is very powerful, in part because it is material and consistent—you have a specific space that you return to, again and again, where you can do the work that you need to do. Physical, material objects have heft, weight, and power, but as discussed in chapter 5, an altar can be many things, and may be energetic rather than physical.

One advantage of maintaining a working transcestral altar, and in cultivating a consistent relationship with transcestors, is that you have a foundation to build on for more specific engagement. For example, if you want to connect with a blood-relative transcestor, you can place an image of them, or something they owned or liked, or really anything that represents them (including a piece of paper with their name on it) on the altar. Now it has become a more general transcestor altar as well as a specific place where you can connect with that person.

If you want to connect with an elevated transcestor, it is much the same process. Place an image of them, or something that represents them or connects you to them, on the altar itself. For working with a specific group of transcestors, it may be a little more esoteric. But, for example, if I wanted to connect to all the transcestors who knew what it was like to take HRT, I would put my used T vials on that altar. If I wanted to connect with all the transcestors who knew what it was like not to have access to gender-affirming care, I would write a poem or create a piece of art that represented my experience, and place that on the altar.

After that, it is much the same as the ritual above. You take the time to call out to them, invite them into your altar and into your awareness. Make offerings to them and make sure that they know you're reaching out to them. One crucial step in this, though, is that it's not a good idea to reach out to a known ancestor if your actions directly contradict their energy. For example, Marsha P. Johnson is a well-known elevated transcestor who many folks reach to and work with (me included). But Marsha was a sex worker and was frequently homeless or unstably housed. She fought hard for the street kids, and the sex workers, and the many who were both, and if you wish to work with her, you should be supporting these groups as well.

This might mean volunteering at a food pantry, donating money and goods to local queer youth organizations, advocating for homeless folks and unstably housed folks, or getting to know the folks who live on the streets around you. You can't disdain Marsha's people while venerating or uplifting Marsha. The energetics of it don't work. It would be like calling in Lou Sullivan and then being hateful to gay men. Know the transcestor you're working with and make sure that you are working in integrity and with consistency.

If you're working with a specific transcestor or group of transcestors, and you're not sure how to give offerings or what to give, I recommend that you take time at your altar and ask them. Then, go take a walk. You might notice a particularly charming leaf, flower, or rock. You might notice a flyer for a certain type of food, or a billboard with a helpful phrase. Just take a little time observing the world around you and see what pops out. Sometimes I will specifically do this at a market or a grocery store. I will ask, "What offerings would you like?" And then purchase them (within reason). If they ask for something outside my comfort zone or budget, I'll just say so and ask what would be an acceptable alternative.

Similar to the previous ritual, this is about relationship building. Showing up when you say you will, how you say you will. Learning not to overpromise and underdeliver. Above all, consistency. If you struggle to do things daily or consistently, let the transcestors know that and see if you can figure out a schedule or option that works for you. You will always do better when you're being realistic about your capacity, needs, and schedule.

Rather than write out any particular ritual, I will offer a few ideas for how you can work with these transcestors, and you can take these ideas and adjust them as needed.

Transcestral Meal

Take the time to cook something for your transcestors. If you happen to know their favorite food, make that (or buy it, but making it is really special). If you don't, then make a dish or meal that is particularly important or special to you. Make sure to put a portion of it on your transcestral altar. You can leave this portion for a bit and then eat it, or you can compost it. Don't put it in the trash and don't let it mold on the altar.

Bespoke Prayer or Poem

Write some sort of spoken invocation, prayer, poem, or similar to the transcestors that you're working with. It's okay if it feels weird or if you don't love your first draft. You can adjust the wording over time. I find that when I write a prayer, it is in the saying of it over and over again that I find its true energy and power. So once you've written this offering, speak it regularly. It's best not to write anything that is too terribly long—you don't want speaking it to be too daunting. Instead, stick with something that you can say in a few minutes, but focus on saying it wholeheartedly.

Creative Practices

Sit before the altar and invite in the specific transcestors you're working with. Bring some sort of creative medium, whether that's pencil and paper, clay, painting supplies, or anything else. Call out to the transcestor and invite them in, then engage with your creative medium. Try to work as intuitively as you can, without concern for how "good" the end result will be. The goal is connection, not perfection.

Throw a Party

This one does require other people, so it may be more specific. But if there's a specific transcestor you want to honor, throw a party in their honor. Be intentional and specific about the activities, the food, and the guests. Make it a rager, a delicious dinner party, a casual potluck—whatever feels energetically aligned for the specific transcestor you're working with. Invite the transcestors to join, and enjoy!

Whatever it is that you choose to do, if you feel it isn't working or isn't effective, try not to fret. There are many ways to connect with specific transcestors, and you can try several to see what works best. But also recall that all of our transcestral relationships rely on consent, just like our relationship with living beings. The transcestors don't have to commune with you, and if you're getting a "no" or a "not right now" from a specific transcestor or group of transcestors, you may consider reaching out to another.

In general, I have found the transcestors to be incredibly receptive and broadly supportive. Reach out with an open heart and in your integrity, and let yourself be delighted and surprised when they reach back.

Asking for Transcestral Guidance or Support

If you make a new friend, it's rare that you would turn around and start asking for favors the next day, right? Working with transcestors, or any other spirit and energetic allies, is similar. Developing a relationship is vital, and that relationship should include reciprocity. That is, you're not just showing up when you want something, and you're not just talking so that you can make requests.

So if you want to make a request or a petition, it's best to start with the first ritual in this chapter, and if you want to work with specific transcestors, it's good to follow the advice in the second ritual. If you're in a desperate situation, you can ask the transcestors directly, but even then, I recommend that you maintain awareness of reciprocity: make offerings, acknowledge that this is a sudden thing you're springing on them.

The other reason to develop a relationship first is so that you can actually recognize or accept the help when it comes to you. If you've never had a conversation with the transcestors before, you might not know or understand how to hear them. You might not yet know what their interventions feel like. So both for the sake of propriety and the sake of your own efficacy, it's wise to develop these relationships mindfully, first and foremost.

But once you've built your altar and tended it, what then? Relationship building and devotional practice can last as long as they need to. In fact, there needn't be any petitions or requests in particular unless you want there to be. But if you do, this ritual is meant to guide you through that process.

There are a few helpful rules for making petitions, and other folks have written about this in more depth than I can here. If you want to go deeper, I recommend Aidan Wachter's book, *Six Ways*, as a great primer. But for our purposes, we're going to work with petitions as a statement of what we are calling into being. The statement should be worded in the present tense, it should be as brief as possible, and it should be written using only positive language.

What I mean by this is that a petition for protection could be worded, "I am safe and well." A petition for wealth could be worded, "Resources flow to me easily, and I have all that I need." A petition for self-love could be, "I adore myself in all my complexity." You can experiment with various phrases, and really think through the implications of the words. Some complex topics might need a little longer petition, and that's okay, but try to practice an economy of language as much as possible.

There are many ways to submit this petition or request to the transcestors. We will be using a candle to submit our petition. Take time planning this ritual and gathering the supplies. You will choose items to "dress" the candle or prepare it for use, so choose your items carefully, using either research or intuition, or both. Choose your candle carefully as well. There are many charts online and in books that discuss which colors are for which types of petitions, but for this ritual we're going to keep it simple, with a white (or natural beeswax) candle. There are also many different shapes, sizes, and types of candles that you can learn all about from other sources.

For this ritual I am recommending whatever candle you can burn safely.

A word on safety: you should never leave a burning candle unattended. If you can't keep an eye on it, extinguish it! If you want to leave it burning while you putter around the house or do other things, place the candle on the stovetop or in the sink, where it won't catch anything else on fire if it falls. Make sure the candle cannot be accessed by pets or children. And finally, if you cannot burn a candle for any reason, there are many other ways to engage in this ritual. I will recommend making a charm bag, where you add the various ingredients into a small fabric bag or pouch. But do whatever works best for you, and as always, feel free to improvise.

Ingredients

- candle, white or natural beeswax, any size you can burn responsibly, and something to light it with
- herbs that align with your petition, or that have personal meaning to you
    - For herbal correspondences, you will likely need to do some research. While some longstanding cultural medicines and traditions have strong associations between particular herbs and their purposes, many modern "wellness" or "alternative spirituality" practices play fast and loose with what different herbs mean and how they function energetically or physically. If you do not have a cultural root you can work with for this, I recommend doing your research, getting to know different herbs through teas, tinctures, and just talking to them, and basically just building your own web of relationships as much as possible. See the Resources section for some of my favorite trans herbalists and herbal resources.

- something to scratch into the wax with: scribe, ballpoint pen, or tack
- anything else that you wish to dress the candle with: glitter, oils, powders
- paper, cut or torn into a square shape, and something to write with
- offerings for your transcestral altar
- If doing the charm bag version of this ritual:
    - a small fabric bag: any color or type of fabric that appeals to you
    - any small objects that align with your petition: rocks, dried flowers, charms, anything that will fit in your small bag
    - optional: a small bit of your hair

Method

1. Begin by doing energetic hygiene to the extent that you need to for this ritual. The intensity of your energetic hygiene should match the intensity of your purpose, but if you're pressed for time and really desperate for assistance, just do what you can do to cleanse, ground yourself, and put up boundaries. The ritual should be done at your transcestral altar and should be conducted with the transcestors that you have been building a relationship with already. Take your time to call in the transcestors that you wish to work with, and name the various offerings that you have brought for them. At minimum, a cup of fresh water should be offered.
2. Say the prayer from the first ritual, or another prayer of your choosing. You may also write a prayer or invocation specifically for this ritual if you'd like.

3. Take your square of paper and write your petition down in normal handwriting, again and again, until you've filled the entire square. Then turn the square so that what was the top is now the side. Again, write your petition down repeatedly until the square is filled. Turn it again, and write again, and continue this until you have written your petition from each side of the square. The words will likely be pretty illegible now, and that is exactly what you want. Now, starting from the top, write your full name, date of birth, and place of birth if you have it. If not, just put as much of that information as you do have, and anything else you can think of that is highly specific to you. Write this again and again until it fills the whole square. Again, turn the paper, and write the information again and again from the next direction, and on, until you've written your identifying information from each side of the square, which at this point will be almost entirely covered with ink and unreadable.
4. If using the candle, take the scribe, ballpoint pen, tack, or other pointed object and scratch your name into the candle on one side, from bottom to top. Then, write your petition in from top to bottom. If using a tea light or glass pillar candle, write your name and petition in a circle around the top. If your petition is too long, write keywords or make a sigil for it instead. Brush off the wax shavings and dress the candle in whatever way you see fit: oils, herbs, glitter, powder, and so on.
5. If using the charm bag, fill it now with everything that you elected for it: herbs, charms, and anything else. Save your hair aside to add last.
6. Light the candle, or take the charm bag and put it before you. Read the petition out loud. If using a candle, drop some of the wax into the petition paper, add a bit of your spit, and fold it

over so the wax sticks the paper together. If using a charm bag, fold your hair into the petition paper and place it inside of the bag.

7. Finish the ritual by thanking the transcestors for their assistance, and sending gratitude to yourself for conducting the ritual. Then, leave the candle and petition or charm bag on the altar to germinate. Try not to look for the answer or solution or blessing that you asked for. It will come when and how it comes, or it won't, and you can do another petition. Working with the transcestors or any spirit being is not like putting money into a snack machine. They are specific, and sometimes we do everything "right" and still don't get the outcome we desired. This is just part of life.
8. No matter what happens, keep making your offerings and checking in with the transcestors. Their presence in your life will absolutely be a blessing, no matter what happens with any specific petition.

Conclusion

I am writing this conclusion from a queer and trans coworking and events space in Koreatown in Los Angeles. Finding this space felt like leaving the world behind and walking into a dream. It is beautiful, whimsical, and affordable as a coworking space in a city where most such spaces are several hundred dollars a month at least. It is the kind of space where, during the day, you might see some of the cutest queers in LA tapping away on their laptops, and in the evening, you might be able to come to a poetry collaging event, community potluck, art gallery opening, or fund raiser for Palestine.

When I talked to the proprietor earlier this week, they said, "People are doing all sorts of crazy things, so I figured, maybe I can just do my thing too."

This idea hit home for me deeply, and echoed so much of my own thinking about becoming a gender doula—a profession that didn't exist when I declared that I was going to leave academia to pursue it so many years ago. There's a sort of gorgeous audacity to being trans, and living into that audacity in one area (identity) can open you up to expressing it in all sorts of other areas of your life.

It reminds me of The Fool card in the tarot. In the traditional decks illustrated by Pamela Colman Smith, The Fool is depicted as a person happily traipsing toward what looks like the edge of a cliff, a little dog dancing up beside them. The person has a little bundle on a stick, much like the ones I used to tie onto sticks when I pretended to run away from home as a kid. They hold a flower in their other hand, head lifted to the heavens, a sunny sky behind them.

The Fool represents a leap of faith. They represent limitless potential, the moment before you move toward some unknown end. To me, The Fool can be a gorgeous representation of the moment when you stop overthinking something and just allow yourself to go toward it wholeheartedly. To everyone around you, it might look like you're stepping off a cliff. But the wisdom of The Fool is knowing that no matter what comes after that step, they have all that they need to fly.

I don't know what the world will be like when you read this, but as I am writing it, it is easy to feel that things are quite bleak. The attacks on trans people have reached a fever pitch, and there's no sign that they are slowing or stopping. The world seems to be hurtling toward multiple disasters, and the term "meta-crisis" has become common in the general parlance. To hold hope, to build community, to explore or transition your gender in these times feels like the ultimate act of audacity.

Yet trans audacity is exactly what we need to meet this moment. This space where I am sitting and sipping an iced coffee and writing my little trans book exists only because of the audacity of one trans person and the support of their community. This book, written and published as fascism rises and transphobia proliferates, exists only because of the audacity of a trans person, a trans editor, and a publisher that refuses to comply in advance. Every single person who chooses to question, explore, or transition their gender is throwing themself off the cliff edge of this society's obsession with control, dominance, and power-over. Those who take that leap find their own ability to fly.

As I've been moving through this time, holding clients through the ups and downs of it, talking to friends and community about it, the thing that I keep coming back to is this: There will be an After. How am I so sure? I've studied trans history, and the backlashes that we've faced have been distressingly cyclical. Every few decades, cis people get up in a tizzy for one reason or another and turn their ire toward trans people. Between these panics, the transphobia generally simmers below the surface, breaking through here and there until the panic whips it into a frenzy and we get a full-blown backlash.

These times are hard. People are hurt; people die. I don't want to bypass that, and I will do everything that I can to minimize the damage. But no matter how severe things get, there is always After. As the current broken systems crumble around us, After is all that I want to think about. Because whatever After looks like, I want it to look like something that actually supports trans people. I want gender exploration to be standard, gender-affirming health care to be accessible, affordable, and high quality, and trans people to be cared for as a matter of course. Whatever After looks like, I want there to be lots and lots of gender rituals.

Taking the time to bring ritual into our own lives is a beautiful thing. If you've done even a few of the rituals in this book, I know that you've seen their power. I've written a lot of these rituals in such a way as to be available for solo use. I know that for many folks, something so tender and so personal may only feel safe in the privacy of solo ritual. Yet, when we are able to do ritual in community, there is a power that echoes out and touches many more folks than we could have imagined. For a negative example, you need only look to gender reveal parties. What started as one random mom's idea has now proliferated into a whole industry, and one that unfortunately seems to involve an unusual amount of death and destruction (probably because it's a cursed enterprise—I'm only half joking).

I'll be honest: I don't want to live in a world where cisgender, heterosexual ideas of gender determine our rituals. The beige box of cishet gender norms can't possibly account for all the color and texture of folks' actual gender. When we free ourselves from the chokehold of these repressive standards, we are able to start stretching into our actual, authentic selves. And when we do that with community, we show those around us a new vision of what is possible. When we open the eyes of those around us to these new vistas of possibility, we are creating the conditions for an After that we actually want to live in.

This book is one of my contributions to the After. I envision a world where it's normal to be invited to someone's renaming ceremony. Where folks of all genders and expressions feel empowered to shift their bodies, not because they hate them, but because they desire an authentic embodiment. I vision a world where transness and queerness and all of it is a place of belonging and joy, of care and justice. And I have the trans audacity to vision all of this even as the world falls apart around us. Even as we are attacked and maligned, even as I am actively coaching folks on how to relocate to a new state or a new country. It is a complex time, and each of us must learn to hold multitudes, lest we be overwhelmed by any specific terrifying reality.

An important aspect of this book is to provide you with some of the tools that you need to survive hard and complex times. The energetic hygiene that I detail in chapter 4, and that I encourage you to do before each ritual, is actually a vital survival medicine. Working with cleansing, grounding, and protection daily or regularly has been fundamental to my mental and physical health, my ability to maintain my hope regardless of these times, and my ability to keep working with After while staying grounded in the now. The other ritual elements I teach in the first part of the book can help you start developing your own relationship to ritual and to magic—two skill sets that have impacted every part of my life, that have changed me on a cellular level.

The rituals in chapter 6, which bring you into deeper trust and relationship with your physical body, are also vital survival medicine. Your body and your intuition hold so much wisdom, and learning how to work with them will help to lead you toward the outcomes that support your thriving. Also, learning to trust your body and to be in dialogue with your body makes it much harder for others to gaslight you about your needs and desires. You do not need to be perfect at hearing your body before you start to shift your gender identity or expression, but the more you can hear your body, the less susceptible you will be to the interference of other people's opinions, anxieties, or baggage.

Finally, the transcestor rituals in chapter 12 create an entry point to one of the most sustaining and supportive spiritual relationships I can imagine. The transcestors called me to be a gender doula, and since I said yes to that calling, everything in my life has changed. While it has been a *tremendous* amount of very hard work, the labor has been joyful and has always pushed me to be real about my boundaries, capacities, and needs. If you don't believe in spiritual relationships or ancestral work, I encourage you to give it a shot and see what happens.

Every part of this book is a love letter to transness, to trans people, and to the magic of gender exploration. My dearest hope is that you, reader, will fall a little more in love with yourself, your community, and your transcestors as you conduct or experience these rituals. No matter what is going on when this book finds its way to you, I most sincerely hope that you are finding the magic of your own authenticity and the bolstering strength of hope.

And above all, always, lean into that trans audacity. The world desperately needs it.

Resources and Further Reading

The practices in this book are informed by many teachers. Some of those teachers are human, and some of them even have resources that you can avail yourself of. This list is inherently incomplete and does not represent everyone or everything that I have learned from. So much of the knowledge I have has come from classes, workshops, and sharing spaces that will never exist in the same way again. Other parts of my knowledge are from my own practice, my relationships with nonhuman beings, transcestors, and ancestors. You can build up a similarly ephemeral but powerful knowledge base through practice and effort. But I have tried to include a good variety of resources to get you started.

Books

adrienne maree brown, *Emergent Strategy: Shaping Change, Changing Worlds*

Aidan Wachter, *Six Ways: Approaches & Entries for Practical Magic*

Alexis Pauline Gumbs, *Undrowned: Black Feminist Lessons from Marine Mammals*

Althaea Sebastiani, *By Rust of Nail & Prick of Thorn: The Theory & Practice of Effective Home Warding*

Cassandra Snow, *Queering the Tarot*

Charlie Claire Burgess, *Radical Tarot: Queer the Cards, Liberate Your Practice, and Create the Future*

Charlie Claire Burgess, *Queer Devotion: Spirituality Beyond the Binary in Myth, Story, and Practice*

Fariha Róisín, *Who Is Wellness For? An Examination of Wellness Culture and Who It Leaves Behind*

Lane Smith, *78 Acts of Liberation: Tarot to Transform Our World*

Lindsay Mack, *Tarot for the Wild Soul: A Trauma-Informed Approach for Growth, Healing, and Grounding in the Present Moment*

Meg Jones Wall, *Tarot Spreads: How to Read Them, Create Them & Revise Them*

Prentis Hemphill, *What It Takes to Heal: How Transforming Ourselves Can Change the World*

Resmaa Menakem, *My Grandmother's Hands: Racialized Trauma and the Pathway to Mending our Hearts and Bodies*

Robin Wall Kimmerer, *Braiding Sweetgrass: Indigenous Wisdom, Scientific Knowledge, and the Teachings of Plants*

Robin Wall Kimmerer, *The Serviceberry: Abundance and Reciprocity in the Natural World*

Sara L. Mastros, *Orphic Hymns Grimoire*

Sarah Faith Gottesdiener, *The Moon Book: Lunar Magic to Change Your Life*

Podcasts

Moonbeaming by Sarah Faith Gottesdiener

The Emerald by Joshua Michael Schrei, especially:

"Inanimate Objects Aren't Inanimate (or Objects)"

"Your Consciousness Comes from the Moon"

Classes

Rowan + Sage online herbalism courses: www.rowanandsage.com

Damien Rourke's Candle Magic Class: www.vellumcodex.com/candle-magic-masterclass

Classes and other resources from the Moon Studio: www.moon-studio.co

Herbalists I Love

Ayelet, Doll Herbalism: www.dollherbalism.com

Laurel Birch, Ms. Tea Botanica: www.msteabotanica.com

Alexis J. Cunningfolk, Worts & Cunning Apothecary: www.wortsandcunning.com

Other Practitioners

Ivy Bromius, an excellent guide on making sigils: www.circlethrice.com

Ursula Rising, astrologer: www.ursularising.com

Diana Rose, astrologer: www.ddamascenaa.com

Helpful Websites

Planetary Hours: www.planetaryhours.net

Sphere + Sundry: www.sphereandsundry.com

Index

C

D

E

F

R

S

About the Author

Photo credit: Abby Mahler

Eli Lawliet, PhD (he/him), is the dreamer and weaver of The Gender Doula. In addition to a decade of experience researching trans health care, law, policy, and history, he has formal training as a full-spectrum doula, breathwork facilitator, and tarot reader as well as many years of surviving the world of poverty wages and retail work. The broad scope of his experience directly informs the spectrum of his work as The Gender Doula.

Eli has been seeing clients since January of 2020 and has worked with hundreds of trans folks nationally and internationally in sessions, classes, and workshops. Though he was born in Oklahoma and raised in Missouri, Eli currently lives in Los Angeles (Gabrieliño-Tongva, Chumash, and Kizh land) with his partner, three cats, a dog, and four snakes.

About North Atlantic Books

North Atlantic Books (NAB) is an independent, nonprofit publisher committed to a bold exploration of the relationships between mind, body, spirit, and nature. Founded in 1974, NAB aims to nurture a holistic view of the arts, sciences, humanities, and healing. To make a donation or to learn more about our books, authors, events, and newsletter, please visit www.northatlanticbooks.com.